Atlantis Studies in Dynamical Systems

Volume 6

The "Atlantis Studies in Dynamical Systems" publishes monographs in the area of dynamical systems, written by leading experts in the field and useful for both students and researchers. Books with a theoretical nature will be published alongside books emphasizing applications.

More information about this series at http://www.springer.com/series/11155

Stephan Mescher

Perturbed Gradient Flow Trees and A_∞-algebra Structures in Morse Cohomology

 Springer

 ATLANTIS PRESS

Stephan Mescher
Mathematisches Institut
Universität Leipzig
Leipzig
Germany

Atlantis Studies in Dynamical Systems
ISBN 978-3-030-09526-0   ISBN 978-3-319-76584-6   (eBook)
https://doi.org/10.1007/978-3-319-76584-6

Mathematics Subject Classification (2010): 58E05, 37D15, 37C10, 57R70

Printed on acid-free paper

This Springer imprint is published by the registered company Springer International Publishing AG
part of Springer Nature
The registered company address is: Gewerbestrasse 11, 6330 Cham, Switzerland

Acknowledgements

During the work on the contents of this book, I was supported by a Ph.D. scholarship of the Deutsche Telekom Foundation and a Ph.D. fellowship of the (meanwhile disbanded) Graduate School "Analysis, geometry and their interactions with the sciences" at the Mathematical Institute of the University of Leipzig.

The contents of this book are part of my Ph.D. thesis [Mes16a], which was submitted to the University of Leipzig and supervised by Matthias Schwarz. I thank Prof. Schwarz for his guidance and support and for raising my interest in A_∞-structures in Morse and Floer theory. Moreover, I am grateful to Alberto Abbondandolo for many helpful discussions and his neverending patience.

I further thank Juan Ojeda, Matti Schneider and Roland Voigt for helpful comments on earlier versions of the manuscript as well as Hendrik Jasnoch and Dirk Scheele for spotting typos and improving the clarity of my sentences.

For his interest in my manuscript and his effort in making the publication of this book possible, I am much obliged to Boris Hasselblatt. I thank Debora Woinke and Zeger Karssen from Atlantis Press for their support in the publication process.

Contents

About the Author

Stephan Mescher (*1984) is a Postdoctoral Researcher at the University of Leipzig. He studied mathematics in Bielefeld and Leipzig and worked at the Ruhr University Bochum and the Queen Mary University of London before. He obtained his diploma in mathematics at Bielefeld University in 2008 and his Ph.D. (Dr. rer. nat.) in 2017 from the University of Leipzig under the supervision of Prof. Dr. Matthias Schwarz. His research interests lie in geometry and topology with a particular focus on Morse-theoretical methods.

Symbols

$(\mathrm{Hess}\,f)_x$	The Hessian bilinear form of f at x
(T_1^e, T_2^e)	Splitting of T into two edges by inserting edges and a vertex at $e \in E_{int}(T)$
$\mathcal{A}_{\mathbf{Y}}^d(x_0, x_1, \ldots, x_d, T)$	The space of $\mathbf{Y}$-perturbed Morse ribbon trees modelled on T starting in x_0 and ending in $(x_1, \ldots, x_d)$
$\#_{\mathrm{alg}}$	The oriented intersection number of an oriented zero-dimensional manifold
$\mathcal{B}_{\mathbf{Y}}((x_0, (y_{e-})_{e \in F}), (x_1, \ldots, x_d, (y_{e+})_{e \in F}), T, F)$	A boundary space of $\mathcal{A}_{\mathbf{Y}}^d(x_0, x_1, \ldots, x_d, T)$ induced by geometric convergence along F
$\mathrm{BinTree}_d$	The set of d-leafed binary trees
$\mathrm{Crit}\,f$	The set of critical points of f
$\ddagger_{r,s}$	Exponents of the signs defining $\mu_{r,s}^*$
δ^g	Morse codifferential of (f, g)
Δ_T	The tree diagonal associated with the tree T
$\varepsilon_\partial(x)$	The boundary orientation of the boundary point x of an oriented one-dimensional manifold
$\maltese_i^j$	Sum of the reduced indices from the $(i+1)$-st to the $(j+1)$-st element of a tensor product $M \otimes A^{\otimes n}$
$\mathcal{M}(f, g)$	The space of finite-length trajectories of $-\nabla^g f$
$\mathcal{M}(x, y, g)$	The space of parametrized Morse trajectories from x to y with respect to g
$\mathcal{M}(Y)$	The space of Y-perturbed finite-length trajectories of $-\nabla^g f$
$\mathcal{M}_{\mathbf{Y}}^d(x_0, x_1, \ldots, x_d, T)$	The space of families of perturbed Morse trajectories modelled on T ending in $(x_1, \ldots, x_d)$
$\mu_{\mathrm{Morse}}(x, f)$	The Morse index of x with respect to f

$\mu^*_{r,s}$	Dual A_∞-bimodule operation of type (r,s)
$\mu^M_{r,s}$	The (r,s)-type multiplication of the A_∞-bimodule M
μ_n	n-th order multiplication of an A_∞-algebra
$\mu_{d,\mathbf{Y}}$	The operation $C^*(f)^{\otimes d} \to C^*(f)$ defined via $a^d_{\mathbf{Y}}(x_0, x_1, \ldots, x_d)$
∂^g	Morse differential of (f,g)
RTree_d	The set of d-leafed ribbon trees
$\sigma(x_0, x_1, \ldots, x_d)$	Exponent of the sign of $a^d_{\mathbf{Y}}(x_0, x_1, \ldots, x_d)$
$\mathrm{split}_k(l, Z)$	The splitting of $Z \in \mathfrak{X}_0(M, k)$ at length l to a product of perturbations
$v_{\mathrm{in}}(e)$	The incoming vertex of the edge e
$v_{\mathrm{out}}(e)$	The outgoing vertex of the edge e
$\widehat{\mathcal{M}}(x, y, g)$	The space of unparametrized Morse trajectories from x to y with respect to g
$\mathcal{W}^s(x, f, g)$	The space of half-trajectories of $-\nabla^g f$ defined on $[0, +\infty)$ ending in x
$\mathcal{W}^u(x, f, g)$	The space of half-trajectories of $-\nabla^g f$ defined on $(-\infty, 0]$ starting in x
$\mathfrak{X}_-(M)$	Perturbation space for negative half-trajectories
$\mathfrak{X}(k_1, k_2, k_3)$	The space of perturbations of type (k_1, k_2, k_3)
$\mathfrak{X}(T)$	The space of perturbations for Morse ribbon trees modelled on T
$\mathfrak{X}_+(M)$	Perturbation space for positive half-trajectories
$\mathfrak{X}_+(M, k)$	The space of perturbations with k real parameters for positive half-trajectories
$\mathfrak{X}_-(M, k)$	The space of perturbations with k real parameters for positive half-trajectories
$\mathfrak{X}_0(M)$	The space of perturbations for finite-length trajectories
$\mathfrak{X}_0(M, k)$	The space of perturbations with k real parameters for finite-length trajectories
$\mathfrak{X}^{\mathrm{back}}_\pm(M, k)$	The space of k-parametrized background perturbations for positive and negative half-trajectories, respectively
$\mathfrak{X}^{\mathrm{back}}(k_1, k_2, k_3)$	The space of background perturbations of type (k_1, k_2, k_3)
$\mathfrak{X}^{\mathrm{back}}(T)$	The space of background perturbations for Morse ribbon trees modelled on T
$\mathfrak{X}^{\mathrm{back}}_0(M, k)$	The space of k-parametrized background perturbations for finite-length trajectories
$a^d_{\mathbf{Y}}(x_0, x_1, \ldots, x_d)$	Algebraic count of elements of $\mathcal{A}^d_{\mathbf{Y}}(x_0, x_1, \ldots, x_d)$
$C^*(f)$	Morse cochain complex of f

$C_*(f)$	Morse chain complex of f
$c_j(\lambda, Y)$	The contraction of Y obtained by inserting λ in its j-th parameter
$k(T)$	The number of internal edges of the tree T
M_{-*}, M^{-*}	Graded module M with inverted grading
$r(T)$	The dexterity of the tree T
T/F	The tree one obtains from T by collapsing all edges in $F \subset E(T)$
$v_0(T)$	The edge of the tree T connected to its root
$v_0(T)$	The root of the tree T
$v_j(T)$	The j-th leaf of the tree T
$v_j(T)$	The edge of the tree T connected to its j-th leaf
$W^+(x, Y)$	The space of Y-perturbed positive half-trajectories ending in x
$W^-(x, Y)$	The space of Y-perturbed negative half-trajectories starting in x
$W^s(x, f, g)$	Stable manifold of x with respect to $-\nabla^g f$
$W^u(x, f, g)$	Unstable manifold of x with respect to $-\nabla^g f$
Y_F	The perturbation induced from Y by geometric convergence along the F-parameters

List of Figures

Introduction

Oh Diane, I almost forgot. Got to find out what kind of trees these are. They're really something.

—Agent Cooper, Twin Peaks

This book's aim is to give a detailed and self-contained exposition of A_∞-algebra structures on Morse cochain complexes of closed manifolds, following works of Kenji Fukaya and Mohammed Abouzaid. We will present a detailed analysis of moduli spaces of perturbed gradient flow trees, which are the crucial objects underlying the A_∞-structures. No previous knowledge on A_∞-algebras is required.

Before we go into detail, we want to give an overview of the main ideas leading to the contents of this book and to provide the reader with the context that they are set within.

Algebraic Topology of Manifolds and Morse Theory

From the beginning of algebraic topology and particularly of singular (co)homology, it has been realized that there are certain topological phenomena occurring in the study of manifolds that do not occur for general topological spaces. Arguably, the most important instance of such a phenomenon is Poincaré duality for closed manifolds and all its related constructions. An example of a more recent result that is characteristic for the topology of manifolds is given by Matthias Kreck and Wilhelm Singhof, who showed in [KS10] that in the category of topological or smooth manifolds, (co)homology theories and cup products can be defined by a set of axioms that is simpler than the Eilenberg–Steenrod axioms.

In particular, it turns out that in categories of manifolds there exist additional algebraic structures in (co)homology. A classical construction is the intersection product on the homology of a closed oriented manifold that is still a useful tool in the study of four-dimensional manifolds. More precisely, the intersection product induces a bilinear form on the second homology group of a closed four-dimensional manifold. Simon Donaldson and Michael Freedman obtained classification results for four-dimensional manifolds by studying their intersection forms, see [Sco05] for an overview.

Another advantage of studying the algebraic topology of smooth manifolds in contrast to general topological spaces is that the singular (co)homology of a manifold possesses an alternative and more analytic description, namely the one of Morse homology. Instead of the singular chain complex, generated by all continuous maps from simplices into a space and thus by uncountably many objects, the Morse chain complex is generated by the critical points of a Morse function on the manifold. Since critical points of Morse functions are isolated, Morse chain complexes have only countably many generators and only finitely many if the manifold is compact. The grading is given by the Morse indices of the points, and the Morse differential is defined by counting unparametrized gradient flow lines between critical points of index difference one. We will explain this construction in a more detailed way in Chap. 1. It is a result usually attributed to René Thom, Stephen Smale and Edward Witten that Morse and singular (co)homology of a smooth manifold are isomorphic. Its original proof is scattered throughout the literature.

The simplicity of Morse (co)chain complexes as rings compared to singular ones is paid by an additional layer of complexity concerning the cup product. While the cup product of a cohomology theory is always associative, in singular cohomology it has the additional property that it is induced by a chain map, which is already associative *on the cochain level*. The cup product in Morse cohomology is induced by an explicit chain map as well, but this map fails to be associative on Morse cochain complexes. Instead, it is part of a more sophisticated algebraic structure, namely the one of an A_∞-algebra, or strongly homotopy associative algebra. The notion of an A_∞-algebra was introduced by James Stasheff in [Sta63b] and motivated by the study of singular chain complexes of based loop spaces. Several decades after Stasheff's seminal work, Kenji Fukaya realized in [Fuk93], see also [Fuk97] and [FO97], that for every closed manifold there exists an A_∞-category whose objects are smooth functions on M and whose morphism spaces are Morse cochain complexes.

A useful modus operandi to understand the topology of manifolds is to mirror classical constructions and results from differential and algebraic topology in Morse homology. For example, Ralph Cohen and Matthias Schwarz gave a Morse-theoretic description of the Thom isomorphism in [CS09], while Roland Voigt described Chern classes and the K-theory of a closed manifold by Morse-theoretic means in [Voi14]. Because of its explicit geometric nature, Morse homology provides a lot of intuition on the geometric background of topological constructions. Morse-theoretic proofs often give a deep understanding of the geometric situation underlying the result. Moreover, Morse-theoretic constructions can give indications of possible generalizations to the infinite-dimensional settings of Floer homology.

In the late 1980s, Andreas Floer established a groundbreaking generalization of Morse homology, nowadays known as Floer homology, for functionals that naturally occur in symplectic geometry and gauge theory, see e.g. [Flo88b, Flo88c,

Flo89a] and [Flo89b]. Floer homology is still a very active area of research. After Floer's seminal works, his methods have been extended to certain functionals in Seiberg–Witten theory, knot theory and contact geometry.

All these flavours of Floer homology have in common that the construction methods of Morse homology have been extended to infinite-dimensional situations, such that finite-dimensional Morse theory can be regarded as a "toy model" for Floer homology. In particular, Morse-theoretic versions of results from topology can give ideas and guidelines for transferring the constructions to Floer homology.

A_∞-Categories of Morse Cochain Complexes

Having realized the singular cohomology of a manifold in terms of Morse theory, it is natural to ask whether the cup product of singular cohomology possesses a Morse-theoretic realization as well. The answer is yes, and such a description of the cup product was discovered by Kenji Fukaya, as well as a family of higher order multiplications on Morse cochain complexes

$$C^*(f_1) \otimes C^*(f_2) \otimes \cdots \otimes C^*(f_d) \to C^*(f_0)$$

for every $d \geq 2$, where $f_0, f_1, \ldots, f_d \in C^\infty(M)$ are Morse functions and M is a closed oriented manifold.

These higher order multiplications had their first appearance in [Fuk93] and were further elaborated on in [Fuk97]. The analytic details were carried out by Fukaya and Yong-Geun Oh in [FO97]. See also [Abo09] for a generalization to compact manifolds with boundary.

A technical difficulty that occurs in their construction is that the multiplications are maps from the product of the Morse cochain complexes of d different Morse functions instead of the d-fold product of the Morse cochain complex of a single Morse function. Here, it is not possible to choose the same Morse function in every factor. To understand this, we have to delve deeper into the definitions of the higher order multiplications.

Similar to the Morse differential and codifferential, the higher order multiplications are defined via counting elements of zero-dimensional moduli spaces. For $d \geq 2$, these moduli spaces consist of continuous maps from a rooted tree (seen as a one-dimensional CW complex) to the manifold M which map the root and the leaves of the tree to fixed critical points, while they edgewise fulfil negative gradient flow equations. For example, let $f_i \in C^\infty(M)$, x_i be a critical point of f_i for $i \in \{0, 1, \ldots, d\}$ and T be a d-leafed rooted binary tree. We consider maps $I : T \to M$ with $I(v_i) = x_i$ for every i, where v_0 denotes the root and v_i denotes the i-th leaf of T for every $i \in \{1, 2, \ldots, d\}$. Moreover, we want that for every edge e there is a Morse function f_e such that

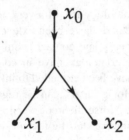

Fig. 1 A configuration used to define the Morse-theoretic cup product

$$\dot{I}_e = -(\nabla^g f_e) \circ I_e$$

where I_e denotes the restriction of I to e. (The edges of T will be parametrized by suitable intervals.) The property of the corresponding moduli spaces being smooth manifolds can be rephrased as a transverse intersection problem. Therefore, the Morse functions involved have to satisfy certain transversality conditions which make it impossible to choose $f_1 = \cdots = f_d$.

Fukaya's method is to start with arbitrary, but distinct, Morse functions associated with every edge leading to a leaf of T. For every other edge, e we define f_e as the sum of the functions $f_{e'}$ for every e' whose incoming vertex coincides with the outgoing vertex of e. For a generic choice of Riemannian metric g, the corresponding moduli space is a smooth manifold. Using these properties, Fukaya was able to define an A_∞-category whose objects are smooth functions on M and whose morphism sets are Morse cochain complexes.

For example, the multiplication for $d = 2$, which is nothing but the chain map inducing the cup product in cohomology, is defined as a map

$$C^*(f_1) \otimes C^*(f_2) \to C^*(f_1 + f_2),$$

if f_1, f_2 and $f_1 + f_2$ are Morse functions on M. The product of the critical points x_1 of f_1 and x_2 of f_2 is defined via counting configurations as illustrated in Fig. 1, where x_0 is a critical point of f_0 of the right Morse index.

Abouzaid's Approach: Perturbing Gradient Flow Lines

In several situations, for example if one is interested in the Hochschild homology of Morse cochain complexes, it is difficult to work with an A_∞-category due to the presence of infinitely many cochain complexes as objects. Instead, it is preferable to work with an A_∞-algebra, i.e. an A_∞-category that has precisely one object, see [Mes16b]. Such a construction is provided by an alternative approach to Fukaya's A_∞-structures in Morse theory, which was suggested by Mohammed Abouzaid in

[Abo11]. Abouzaid defines products $C^*(f)^{\otimes d} \to C^*(f)$ for a closed oriented manifold M, a fixed Morse function $f : M \to \mathbb{R}$ and every $d \geq 2$ by considering moduli spaces of continuous maps from a tree to the manifold, which are trajectories of *perturbations* of the negative gradient flow of a Morse function on a closed manifold. Using transversality theory, one shows that for generic choices of perturbations, the resulting moduli spaces will be manifolds. Then, one can construct the higher order multiplications along the same lines as in Fukaya's works by counting elements of zero-dimensional moduli spaces.

Abouzaid has shown that for good choices of perturbations, the Morse cochain complex $C^*(f)$ with the resulting higher order multiplications becomes an A_∞-algebra. Moreover, his results imply that there is an equivalence of A_∞-algebras between this A_∞-algebra and the singular cochain complex of the space, seen as a differential graded algebra with respect to the singular cup product.

In fact, Abouzaid's perturbation method is the Morse-theoretic analogue of the more sophisticated Floer-theoretic method constructed by Paul Seidel in [Sei08, Part II]. Seidel defined perturbations for pseudo-holomorphic curves which are modelled on boundary-pointed discs, which require more subtle analytic methods, especially concerning compactness properties of the resulting moduli spaces.

An Application: Morse Cochain Complexes and Free Loop Spaces

The author's initial motivation for considering A_∞-structures on Morse cochain complexes is a curiosity for the extent of the topology of a closed manifold that is encoded in gradient dynamics of Morse functions. For this purpose, the author studied the Hochschild homology and cohomology of Morse cochain complexes in his Ph.D. thesis [Mes16a].

Hochschild homology and cohomology are algebraic invariants that are classically associated with associative or differential graded algebras, see [Lod98, Chap. 1] or [Wei94, Chap. 9]. The definition of Hochschild homology and cohomology can be generalized in a straightforward way to the setting of A_∞-algebras, see [Mes16b] for an overview.

By a theorem of Ralph Cohen and John Jones, see [CJ02], the homology of the free loop space of a simply connected closed oriented manifold M is isomorphic to the Hochschild cohomology of $C^*(M)$, the singular cochain algebra of M, both considered with integer coefficients. The theorem of Cohen and Jones is based on an older result of Jones from [Jon87]. Cohen and Jones further showed that this is an isomorphism of rings with respect to the Hochschild cup product and the Chas–Sullivan loop product, see [CS99] or [CV06, Sect. 1.2]. As a fascinating consequence of this result, the homology of the free loop space of a simply connected closed oriented manifold M can be computed without considering loops in M at all, including the Chas–Sullivan loop product.

By a result of Abouzaid from [Abo11, Sect. 3], the Hochschild cohomology of the Morse cochain complex of a Morse function on M, seen as an A_∞-algebra, is isomorphic to the Hochschild cohomology of $C^*(M)$. Combining this observation with the theorem of Cohen and Jones shows that *an A_∞-algebra structure of the Morse cochain complex of a single Morse function on a simply connected closed oriented manifold determines the homology of the free loop space of M.*

In [Mes16a], the author undertook several steps towards a more explicit geometric description of this correspondence. More precisely, he constructed a chain map between the Hochschild cochain complex of a Morse cochain complex on M and a Morse chain complex of the free loop space of M, which is compatible with Morse-theoretic versions of the above-mentioned products and which conjecturally induces an isomorphism in (co)homology for simply connected M.

An Outlook on Fukaya Categories in Symplectic Geometry

Fukaya obtained his original definition of A_∞-categories in Morse theory as a by-product of his considerations of A_∞-categories in Lagrangian Floer theory. The latter are nowadays known as Fukaya categories and are a vibrant research topic in symplectic geometry. To give the reader an impression of the power of A_∞-structures on Morse-type cochain complexes, we give a brief sketch of the construction of Fukaya categories in symplectic geometry and some of their applications.

In the late 1980s, Andreas Floer constructed Morse-type homology groups for Hamiltonian action functionals on free loop spaces of symplectic manifolds, see, e.g., his seminal works [Flo88b, Flo88c, Flo89a] and [Flo89b]. His constructions eventually led to striking results on Hamiltonian dynamical systems, culminating in a proof of Arnold's conjecture on the connection between periodic orbits of Hamiltonian dynamical systems and the topology of the underlying symplectic manifold, see [FO99] and [LT98].

All of these first results have in common that they are obtained by using the topology and geometry of a symplectic manifold to derive results on Hamiltonian dynamical systems. However, shortly after the work of Floer, experts in the field started to reverse this viewpoint. Considering Floer homology as an algebraic invariant of symplectic manifolds, it is unknown to this day to which extent it can be seen as an abstractly defined homology theory. Certain results towards such a notion for a symplectic cobordism category were worked out by Kai Cieliebak and Alexandru Oancea in [CO15].

Fukaya categories are Floer-theoretic invariants of symplectic manifolds. They are based on Lagrangian Floer homology, which was introduced by Floer in [Flo88a] as the homology of chain complexes generated by intersection points of Lagrangian submanifolds of symplectic manifolds or, more generally, Hamiltonian chords connecting Lagrangian submanifolds. The Fukaya category of a symplectic manifold is an A_∞-category whose objects are the Lagrangian submanifolds of the manifold equipped with certain extra pieces of information regarding orientations and gradings and whose morphism spaces are Lagrangian Floer cochain complexes, see [Aur14] for an overview and [Sei08, Part II] for an extensive treatment.

In relation to the contents of this book, a connection between Fukaya's A_∞-categories of symplectic manifolds and A_∞-categories associated with Morse cochain complexes is given by considering cotangent bundles. The cotangent bundle of a closed manifold can in a canonical way be equipped with the structure of a symplectic manifold. From a physics point of view, the cotangent bundle of a manifold M is interpreted as the phase space of a particle moving in the space M.

As explained by Fukaya and Yong-Geun Oh in [FO97] and by Fukaya in [Fuk97], Fukaya categories of cotangent bundles encode phenomena from quantum physics in this physical interpretation. As discussed in the given references, A_∞-structures on Morse cochain complexes may be seen as *classical limits* of A_∞-structures in cotangent bundles.

While one might guess that Fukaya categories classify symplectic manifolds completely, a confirmation or a falsification seems far out of reach of current research. Even the question of the well-definedness of the Fukaya category leads to very delicate analytic difficulties, see [FOOO09] for a detailed discussion of these aspects.

Nevertheless, Fukaya's A_∞-categories have proven to be tremendously useful in symplectic geometry. Most notably, they play an important role in mirror symmetry, an important conjectural relation between symplectic geometry and complex geometry. At the heart of mirror symmetry lies a conjectured duality between Calabi–Yau manifolds which occur in "mirror pairs", for which the symplectic geometry of the one determines the complex geometry of the other and vice versa. In his talk at the ICM 1994, Maxim Kontsevich reformulated mirror symmetry as an equivalence of algebraic invariants, namely of A_∞-categories, see [Kon95]. Kontsevich's formulation is known as the homological mirror symmetry conjecture. On the symplectic side, Kontsevich considers derived Fukaya categories which are supposed to be dual to derived categories of coherent sheaves on the complex side. While we will not go into further detail, this conjectured correspondence shows that Fukaya categories are supposed to contain a tremendous amount of information about symplectic manifolds.

Besides homological mirror symmetry, Fukaya categories have already found various applications in symplectic geometry. Recently, Abouzaid and Thomas Kragh used Fukaya categories to show in [AK16] that two lens spaces are diffeomorphic if and only if their cotangent bundles are symplectomorphic. Lens spaces are classic examples of spaces which are homotopy-equivalent, but non-homeomorphic. They are commonly used as test objects to check whether an invariant recognizes more than the homotopy type of a topological space. Using Fukaya categories, Abouzaid further showed in [Abo12] that the symplectic structures of their cotangent bundles may be used to distinguish exotic differentiable structures on $(4k+1)$-spheres, where $k \in \mathbb{N}$.

Overview of the Book

The aim of this book is to give a detailed construction of the higher order multiplications on the Morse cochains of a fixed Morse function using the perturbation methods of Abouzaid. Moreover, we will show that for convenient choices of perturbations the Morse cochain complex equipped with these multiplications becomes an A_∞-algebra. We will provide the reader with many analytic details and a self-contained construction of higher order multiplications along Abouzaid's ideas.

The basic ideas of this book are implicitly contained in [Abo11]. Many of the results are stated or partially proven in Abouzaid's work and we will provide the analytic details underlying Abouzaid's methods. For the sake of readability, we refrain from giving a reference to the corresponding result in Abouzaid's work at every result we are considering and see this paragraph as a general reference to [Abo11].

We assume that the reader is familiar with standard results of Morse theory. The classic reference for Morse theory is [Mil63]. We will loosely orient our treatment of Morse homology on the textbook [Sch93]. In Chap. 1, we will provide further references on Morse theory and Morse homology.

Throughout this book and especially in Chap. 2, we will use several results for spaces of (unperturbed) negative gradient flow trajectories and give references to those results whenever they are needed. Moreover, we will rely on certain gluing analysis results for Morse trajectories which transfer with only minor modifications to our situation. The necessary results are contained in [Sch93, Sch99] and [Weh12]. Furthermore, Chaps. 4–6 require certain results from graph theory which we will mostly state without providing proofs. The proofs of those results are elementary and therefore left to the reader.

The main part of this book starts with a concise introduction to Morse homology for finite-dimensional manifolds in Chap. 1. In contrast to the other chapters, we will not prove the main results and give references to the literature instead.

In Chap. 2, we carefully introduce the necessary spaces of perturbations and the different kinds of perturbed negative gradient flow trajectories. We will rely on certain results on unperturbed negative gradient flow trajectories which we will briefly present. Thereupon, we discuss moduli spaces of these perturbed trajectories and certain evaluation maps defined on them.

The results about the evaluation maps are further generalized in Chap. 3. After introducing certain more general perturbation spaces, we consider moduli spaces of families of perturbed Morse trajectories satisfying endpoint conditions, which are in a certain sense nonlocal. Finally, we state and prove a general regularity theorem about these moduli spaces.

This nonlocal regularity theorem is applied to a special situation in Chap. 4. This section starts with the introduction of several notions from graph theory. Afterwards, we discuss how the results from Chap. 3 can be applied to consider moduli spaces of maps from a tree to a manifold which edgewise fulfil perturbed negative gradient flow equations.

In Chap. 5, we investigate sequential compactness properties of the moduli spaces from Chap. 4. We introduce a notion of geometric convergence with which we can describe the possible limiting behaviour of sequences in the moduli spaces of perturbed negative gradient flow trees. We further give explicit descriptions of the moduli spaces in which the respective limits are lying.

The objects of study in Chap. 6 are zero- and one-dimensional moduli spaces of perturbed negative gradient flow trees. By applying the results of Chap. 5 to zero-dimensional spaces, we are able to define multiplications on Morse cochain complexes via counting elements of these moduli spaces. Applying the above-mentioned gluing results from Morse theory, we will then describe compactifications of one-dimensional moduli spaces and finally use the counts of their boundary components to prove that the Morse cochain complex equipped with the multiplications becomes an A_∞-algebra.

Finally, in Chap. 7, we dualize the A_∞-algebra structures from the previous chapter to construct A_∞-bimodule structures on Morse chain complexes. We first present some general algebraic results before we apply the algebra to the Morse-theoretic setting.

A discussion of orientability and orientations of the moduli spaces of perturbed Morse ribbon trees is required in order to construct higher order multiplications with integer coefficients, since they will be constructed using oriented intersection numbers. For the benefit of the flow of reading, all results about orientations and most necessary sign computations are deferred to Appendix A. We will give references to the respective parts of this appendix whenever required. The reader is advised to ignore questions of orientations upon first reading.

Notational Remarks

If $f : X \to Y$ is a differentiable map between differentiable manifolds, we denote the differential of f at $x \in X$ by $Df_x : T_x X \to T_{f(x)} Y$. Moreover, we let $Df_x[v]$ denote the application of Df_x to $v \in T_x X$. The differential of a differentiable *function* $g : X \to \mathbb{R}$ at $x \in X$ will also be denoted by $dg_x : T_x X \to \mathbb{R}$.

The cardinality of a set S will be denoted by $|S|$.

We call both a map between chain complexes that respects the differentials and a map between cochain complexes that respects the codifferentials a *chain map*. A *quasi-isomorphism* is a chain map between two (co)chain complexes which induces an isomorphism of (co)homology groups.

We further use the convention that the empty set is a manifold of *every* dimension.

Chapter 1
Basics on Morse Homology

This brief chapter is intended to provide the reader with an overview of the construction of Morse (co)homology for finite-dimensional manifolds. We present the main notions and results in a concise way and give references to detailed presentations and proofs whenever appropriate. Moreover, we establish some notation that will be employed throughout this book. Except for notational conventions, a reader familiar with Morse homology might skip this chapter without disadvantages. There are several detailed and recommendable references on Morse homology, see e.g. the textbooks [Sch93, BH04], [Jos08, Chap. 7], [Nic11] or [AD14] as well as the set of lecture notes [Hut02] and the article [Web06].

We assume that the reader is familiar with the basic notions of Morse theory. Together with details and examples, these are contained in the classic textbook [Mil63] as well as in the more recent textbooks [Mat02, Nic11].

In the following, let X be a finite-dimensional smooth manifold. For $k \in \mathbb{N}_0$ we write $C^k(X) := C^k(X, \mathbb{R})$ for the space of k-times differentiable real functions on X. For $f \in C^1(X)$ we let

$$\mathrm{Crit}\ f := \{x \in X \mid df_x = 0\}$$

denote the set of critical points of f. We start our overview by introducing the basic notions of Morse theory.

Definition 1.1 A function $f \in C^2(X)$ will be called a *Morse function* if the Hessian bilinear form $(\mathrm{Hess}\ f)_x : T_x X \times T_x X \to \mathbb{R}$ is non-degenerate at every $x \in \mathrm{Crit}\ f$.

Remark 1.2 The non-degeneracy condition particularly implies that critical points of Morse functions are isolated. In particular, a Morse function on a compact manifold has only finitely many critical points.

© Springer International Publishing AG, part of Springer Nature 2018
S. Mescher, *Perturbed Gradient Flow Trees and A∞-algebra Structures in Morse Cohomology*, Atlantis Studies in Dynamical Systems 6,
https://doi.org/10.1007/978-3-319-76584-6_1

If f is a Morse function, then for every $x \in \text{Crit } f$ the tangent space $T_x X$ possesses a decomposition

$$T_x X = E_x^- \oplus E_x^+ \,,$$

such that $(\text{Hess } f)_x$ is negative definite on E_x^- and positive definite on E_x^+. While these spaces are not unique, it follows from the symmetry of the Hessian and standard results from linear algebra that their dimensions are uniquely determined.

Definition 1.3 Let $f \in C^2(X)$ be a Morse function and $x \in \text{Crit } f$. In terms of the above decomposition of $T_x X$ we call the number

$$\mu_{\text{Morse}}(x, f) := \dim E_x^-$$

the *Morse index of x with respect to f*.

The main idea behind Morse theory is that the study of a differentiable function on a manifold and its critical points yields results on the topology of the manifold. Classically, as it is beautifully presented in [Mil63], one recovers the homotopy type of the manifold as the homotopy type of a CW complex whose cells of dimension k are in correspondence with the critical points of Morse index k. A first direct relation between the critical points of a Morse function and the singular homology of the manifold is given by the Morse inequalities, see [Mil63, Chap. 5] or [Nic11, Sect. 2.3].

In the last decades, the focus of Morse theory has shifted from considering only the critical points of a function along with their Morse indices to the study of the whole gradient flow of a Morse function. We next discuss the relevant notions for this viewpoint.

In the following, we let (X, g) be a Riemannian manifold and $f \in C^2(X)$. We recall that the gradient of f with respect to g is the vector field $\nabla^g f : X \to TX$ that satisfies

$$df_x[v] = g_x(\nabla^g f(x), v) \qquad \forall x \in X, \ v \in T_x X \,.$$

Note that $\text{Crit } f$ is precisely the set of singular points of the gradient of f with respect to any given metric. In the following, we are mainly interested in flow lines of the negative gradient of f.

Definition 1.4 A curve $\gamma \in C^1(\mathbb{R}, X)$ that satisfies

$$\dot{\gamma}(s) = -\nabla^g f(\gamma(s))$$

for every $s \in \mathbb{R}$ is called a *Morse trajectory* of f with respect to g.

One checks that when the following limits exist in X, then

$$\lim_{s \to -\infty} \gamma(s), \ \lim_{s \to +\infty} \gamma(s) \in \text{Crit } f$$

and

$$\lim_{s \to -\infty} \gamma(s) = \lim_{s \to +\infty} \gamma(s) \quad \Leftrightarrow \quad \gamma \text{ is constant.}$$

Definition 1.5 Let $x, y \in \text{Crit } f$. We call

$$\mathcal{M}(x, y, g) := \left\{ \gamma : \mathbb{R} \xrightarrow{C^1} X \;\middle|\; \dot{\gamma} = -(\nabla^g f) \circ \gamma, \; \lim_{s \to -\infty} \gamma(s) = x, \; \lim_{s \to +\infty} \gamma(s) = y \right\}$$

the space of (parametrized) Morse trajectories from x to y. We consider $\mathbb{R}$ as an additive group and the group action

$$\mathbb{R} \times \mathcal{M}(x, y, g) \to \mathcal{M}(x, y, g), \quad (\tau, \gamma) \mapsto (s \mapsto \gamma(s + \tau)) .$$

The quotient of this action

$$\widehat{\mathcal{M}}(x, y, g) := \mathcal{M}(x, y, g)/\mathbb{R}$$

is called *the space of unparametrized Morse trajectories from x to y.*

Remark 1.6 If $f \in C^k(X)$ for some $k \in \mathbb{N}$, then every flow line of $-\nabla^g f$ is of class C^k as well. In particular, if $f \in C^\infty(X)$, then every negative gradient flow line is smooth. This follows from standard results for solutions or ordinary differential equations.

Until further notice, we choose and fix a Morse function $f : X \to \mathbb{R}$ and put $\mu(x) := \mu_{\text{Morse}}(x, f)$ for every $x \in \text{Crit } f$.

We will further assume that $f \in C^\infty(X)$. This condition can be weakened to derive slightly weaker regularity results than the following one, but we refrain from giving the details. The following crucial result was obtained by Stephen Smale in [Sma61].

Theorem 1.7 *For a generic choice of Riemannian metric g on M, the space $\mathcal{M}(x, y, g)$ is a smooth manifold of dimension*

$$\dim \mathcal{M}(x, y, g) = \mu(x) - \mu(y)$$

for all $x, y \in \text{Crit } f$. If $x \neq y$, then the $\mathbb{R}$-action on $\mathcal{M}(x, y, g)$ will be smooth, free and proper, such that in this case $\widehat{\mathcal{M}}(x, y, g)$ is a smooth manifold with

$$\dim \widehat{\mathcal{M}}(x, y, g) = \mu(x) - \mu(y) - 1 .$$

Definition 1.8 We call (f, g) a *Morse–Smale pair* if

- $f \in C^\infty(X)$ is a Morse function,
- f is bounded from below,
- g is a smooth Riemannian metric on X which is generically chosen with respect to f in the sense of Theorem 1.7.

The main idea behind the proof of Theorem 1.7 is to describe $\mathcal{M}(x, y, g)$ as the intersection of a map with a submanifold. The Riemannian metric g is seen as an auxiliary parameter of this map. By equipping a convenient set of metrics with the topology of a separable Banach space, one applies the Sard–Smale transversality theorem to obtain that for generic choice of the metric parameter, the intersection is transverse, implying the regularity result. See [Sch93, Sect. 2.3] for details.

It is crucial in the construction of Morse homology that the spaces $\widehat{\mathcal{M}}(x, y, g)$ have certain sequential compactness properties. To ensure these properties, the Morse–Smale pair has to satisfy an additional condition.

Definition 1.9 Let $f \in C^\infty(X)$ and g be a Riemannian metric on X. The pair (f, g) satisfies *the Palais–Smale condition* if every sequence $\{x_n\}_{n\in\mathbb{N}}$ in X for which

- $\{|f(x_n)|\}_{n\in\mathbb{N}}$ is bounded,
- $\lim_{n\to\infty} \|\nabla^g f(x_n)\|_{x_n} = 0$,

has a convergent subsequence. Here, $\|\cdot\|_x : T_x X \to \mathbb{R}$ denotes the norm induced by g for every $x \in X$.

Remark 1.10 One checks without difficulties that a Morse–Smale pair (f, g) satisfies the Palais–Smale condition if all closed sublevel sets $f^{-1}((-\infty, a])$, $a \in \mathbb{R}$, are compact. In particular, every Morse–Smale pair on a compact manifold satisfies the Palais–Smale condition.

Theorem 1.11 *Assume that (f, g) is a Morse–Smale pair which satisfies the Palais–Smale condition. If $x, y \in \mathrm{Crit}\ f$ satisfy $\mu(x) = \mu(y) + 1$, then $\widehat{\mathcal{M}}(x, y, g)$ is a finite set.*

Another important observation is that for a Morse–Smale pair (f, g) the spaces $\widehat{\mathcal{M}}(x, y, g)$ are *orientable* manifolds. To proceed, we need to equip these manifolds with orientations. In this book, we will only consider the simplest case, namely the one that X is itself orientable. As shown in [Sch93, Chap. 3], the following considerations may be extended to nonorientable manifolds using the framework of coherent orientations that was introduced by Andreas Floer and Helmut Hofer in [FH93].

Assume that the manifold X is oriented and choose and fix an orientation of the unstable manifold $W^u(x, -\nabla^g f)$ for every $x \in \mathrm{Crit}\ f$. (These spaces will be introduced in Sect. 2.1.) Together, the orientation on X and the orientations of the unstable manifolds induce orientations on the stable manifolds of the critical points of f, see Appendix A.1 for details. The spaces $\mathcal{M}(x, y, g)$ are identified with transverse intersections of unstable and stable manifolds, hence the orientations of these spaces induce orientations of the $\mathcal{M}(x, y, g)$. Finally, using the differential of the reparametrization action, this construction yields orientations of the spaces $\widehat{\mathcal{M}}(x, y, g)$.

Throughout the remainder of this chapter, we assume that all Morse functions on oriented manifolds come equipped with fixed choices of orientations of their unstable manifolds and let all spaces of the form $\widehat{\mathcal{M}}(x, y, g)$ be equipped with the orientations we previously described.

Definition 1.12 Let X be a complete oriented manifold and $f \in C^\infty(X)$ be a Morse function.

(a) For every $j \in \mathbb{N}_0$ we define

$$C^j(f) = \prod_{\substack{x \in \text{Crit } f \\ \mu(x)=j}} \mathbb{Z} \cdot x \quad \text{and} \quad C_j(f) = \bigoplus_{\substack{x \in \text{Crit } f \\ \mu(x)=j}} \mathbb{Z} \cdot x$$

and put $C^*(f) := \bigoplus_{j \in \mathbb{N}_0} C^j(f)$ and $C_*(f) := \bigoplus_{j \in \mathbb{N}_0} C_j(f)$.

(b) Let g be a Riemannian metric on X, such that (f, g) is a Morse–Smale pair and let $x, y \in \text{Crit } f$ satisfy $\mu(x) = \mu(y) + 1$. Define

$$n(x, y, g) := \#_{\text{or}} \widehat{\mathcal{M}}(x, y, g) = \sum_{\gamma \in \widehat{\mathcal{M}}(x,y,g)} \epsilon(\gamma) \,,$$

where $\epsilon : \widehat{\mathcal{M}}(x, y, g) \to \{-1, 1\}$ is the function associated with the above mentioned orientation.

(c) For a Morse–Smale pair (f, g) on X we define a graded homomorphism

$$\partial^g : C_*(f) \to C_{*-1}(f)$$

as the $\mathbb{Z}$-linear extension of

$$\partial^g(x) = \sum_{\substack{y \in \text{Crit } f \\ \mu(y)=\mu(x)-1}} n(x, y, g) \cdot y \quad \forall x \in \text{Crit } f \,.$$

Similarly, we define a graded homomorphism

$$\delta^g : C^*(f) \to C^{*+1}(f)$$

as the $\mathbb{Z}$-linear extension of

$$\delta^g(x) = \sum_{\substack{z \in \text{Crit } f \\ \mu(z)=\mu(x)+1}} n(z, x, g) \cdot z \quad \forall x \in \text{Crit } f \,,$$

where the sum is meant to be a formal one.

In particular, $C_j(f) = C^j(f) = \{0\}$ whenever $j > \dim X$. Moreover, $|n(x, y, g)| < +\infty$ for all x and y of the right indices if (f, g) satisfies the Palais–Smale condition, so that ∂^g and δ^g are well-defined.

Theorem 1.13 *Let X be a complete manifold. If (f, g) is a Morse–Smale pair on X that satisfies the Palais–Smale condition, then $(C_*(f), \partial^g)$ is a chain complex and $(C^*(f), \delta^g)$ is a cochain complex, i.e.*

$$\partial^g \circ \partial^g = 0 \quad and \quad \delta^g \circ \delta^g = 0 \, .$$

This theorem is shown by a useful cobordism argument. One considers one-dimensional moduli spaces $\widehat{\mathcal{M}}(x, y, g)$ and investigates their sequential compactness. Here, the Palais–Smale condition is crucial. One shows that one-dimensional spaces of unaparametrized Morse trajectories can be compactified to compact one-dimensional manifolds, whose boundaries are identified with products of zero-dimensional spaces of unparametrized Morse trajectories. The oriented counts of elements of these products are the coefficients of the compositions $\partial^g \circ \partial^g$ and $\delta^g \circ \delta^g$. One derives a relation for these coefficients from this boundary description, eventually yielding that $\partial^g \circ \partial^g = 0$ and $\delta^g \circ \delta^g = 0$. See [Sch93, Sect. 2.4] for details.

Definition 1.14 The chain complex $(C_*(f), \partial^g)$ is called the *Morse chain complex* of (f, g). Its homology

$$H_*(f, g) := H_*(C_*(f), \partial^g)$$

is called the *Morse homology of* (f, g). Analogously, the cochain complex $(C^*(f), \delta^g)$ is called the *Morse cochain complex of* (f, g) and its cohomology

$$H^*(f, g) := H^*(C^*(f), \delta^g)$$

is called the *Morse cohomology of* (f, g).

Theorem 1.15 (Thom–Smale–Witten) *Let (f, g) be a Morse–Smale pair on X that satisfies the Palais–Smale condition. Then there are isomorphisms*

$$H_*(f, g) \cong H_*(X; \mathbb{Z}) \quad and \quad H^*(f, g) \cong H^*(X; \mathbb{Z}) \, ,$$

where $H_*(X; \mathbb{Z})$ and $H^*(X; \mathbb{Z})$ denote the singular (co)homology of X with integer coefficients.

The original proof of this theorem is scattered throughout the literature. For complete proofs see [BH04, Chap. 7] or [Sch93].

Remark 1.16 For Morse–Smale pairs on arbitrary complete manifolds, orientable or not, there exists the analogous construction of Morse (co)homology with $\mathbb{Z}_2$-coefficients. Here, one replaces $C_*(f)$ by the free $\mathbb{Z}_2$-module generated by Crit f and the coefficients $n(x, y, g)$ by the parities of $|\widehat{\mathcal{M}}(x, y, g)|$, seen as elements of $\mathbb{Z}_2$. Analogously, one replaces $C^*(f)$ by the corresponding free $\mathbb{Z}_2$-module. As the reader might expect, the (co)homology of these complexes is isomorphic to the singular (co)homology of X with $\mathbb{Z}_2$-coefficients.

Following the establishment of Morse (co)homology, several authors have discussed how other well-known constructions from the topology of manifolds can be obtained by purely Morse-theoretic means. For example, there are Morse-theoretic constructions and proofs of

- the Eilenberg-Steenrod axioms and Poincaré duality in [Sch93],
- the cup product in [Fuk93, Fuk97],
- equivariant cohomology, see [AB95],
- the Thom isomorphism and umkehr maps in [CS09],
- Steenrod squares and Stiefel-Whitney classes in [CN12],
- the Leray-Serre spectral sequence in [Sch12],
- Chern classes and K-theory in [Voi14].

These results underline the significance and the enormous amount of information that a gradient dynamical system encodes about the topology of the manifold it is defined on.

Remark 1.17 There exist generalizations of the construction of Morse homology to infinite-dimensional Hilbert manifolds, i.e. Banach manifolds which are locally modelled on Hilbert spaces. In [AM06], Alberto Abbondandolo and Pietro Majer present a very general setting for Morse homology on Hilbert manifolds, see also [AS10, Appendix A] for an overview. The upshot is that Morse chain and cochain complexes for Morse functions on Hilbert manifolds can be defined if the Morse function satisfies certain additional conditions which are automatically satisfied in a finite-dimensional setting.

Chapter 2
Perturbations of Gradient Flow Trajectories

As we have mentioned in the introduction, a crucial step towards defining an A_∞-algebra structure on the Morse cochain complex of a single Morse function is to consider *perturbed* Morse trajectories. More precisely, we want to dicuss curves which do not satisfy a negative gradient flow equation, but a *perturbed* negative gradient flow equation of the form

$$\dot{\gamma}(s) + \nabla^g f \circ \gamma(s) + Z(s, \gamma(s)) = 0,$$

where we pick up the notation from Chap. 1 and where Z is a suitable time-dependent vector field. The time-dependence of Z has the important consequence that we may control on which parts of the domain interval of γ, the vector field Z vanishes. In fact, we will choose the perturbations in such a way that they are only non-vanishing close to the ends of the interval.

This chapter's aim is to introduce a precise setting for perturbed Morse trajectories and to discuss properties of their moduli spaces. We will investigate the regularities of these spaces is as well as first results on transverse intersections.

In several aspects of Morse homology, it is useful to identify elements of unstable and stable manifolds of critical points with the trajectories that connect these points to the corresponding critical point. Hereby, elements of unstable manifolds correspond to Morse trajectories defined on $(-\infty, 0]$ while elements of stable manifolds are seen as Morse trajectories on $[0, +\infty)$.

For our purpose, spaces of perturbed analogues of these trajectories defined on semi-infinite are important in the construction of A_∞-structures. Additionally, perturbed Morse trajectories defined on compact intervals play an important role for us. Roughly, we will use them in later chapters to model internal edges of perturbed gradient flow trees. Hence, there are three types of perturbed trajectory spaces that are of relevance for us and we will investigate them one by one in the course of this chapter.

© Springer International Publishing AG, part of Springer Nature 2018   9
S. Mescher, *Perturbed Gradient Flow Trees and A∞-algebra Structures in Morse Cohomology*, Atlantis Studies in Dynamical Systems 6,
https://doi.org/10.1007/978-3-319-76584-6_2

We briefly discuss Morse trajectories defined on semi-infinite intervals and their moduli spaces in Sect. 2.1. Afterwards, we introduce the notion of perturbed Morse trajectories with domain $(-\infty, 0]$ in Sect. 2.2 and with domain $[0, +\infty)$ in Sect. 2.3. In Sect. 2.4 we discuss Morse trajectories on finite-length intervals before introducing their perturbed analogues.

Throughout the remainder of this book, we let M be a closed oriented manifold and $f \in C^\infty(M)$ be a Morse function. We further put $n := \dim M$ and denote the Morse index of any $x \in \mathrm{Crit}\, f$ by $\mu(x) := \mu_{\mathrm{Morse}}(x, f)$.

2.1 Spaces of Semi-infinite Morse Trajectories

We start by recalling some necessary definitions and results on unstable and stable manifolds and their associated trajectory spaces. The former are discussed in the given references on Morse theory, while the latter are found in [Sch99].

Definition 2.1 Let $x \in \mathrm{Crit}\, f$. For a Riemannian metric g on M, the *unstable manifold of x with respect to (f, g)* is defined by

$$W^u(x, f, g) := \left\{ y \in M \;\middle|\; \lim_{t \to -\infty} \phi_t^{-\nabla^g f}(y) = x \right\},$$

where $\phi^{-\nabla^g f}$ denotes the flow of the vector field $-\nabla^g f$, i.e. the negative gradient flow of f with respect to g.
The *stable manifold of x with respect to (f, g)* is defined by

$$W^s(x, f, g) := \left\{ y \in M \;\middle|\; \lim_{t \to +\infty} \phi_t^{-\nabla^g f}(y) = x \right\}.$$

It is well-known from Morse theory, see [Jos08, Sect. 6.3] or [Web06, Theorem 2.7], that for every $x \in \mathrm{Crit}\, f$ the unstable and stable manifolds of x with respect to (f, g) are smooth manifolds of dimension

$$\dim W^u(x, f, g) = \mu(x), \quad \dim W^s(x, f, g) = n - \mu(x),$$

respectively, and that both $W^u(x, f, g)$ and $W^s(x, f, g)$ are embedded submanifolds of M. Moreover, M can be decomposed into the unstable and stable manifolds with respect to (f, g):

$$M = \bigcup_{x \in \mathrm{Crit}\, f} W^u(x, f, g) = \bigcup_{x \in \mathrm{Crit}\, f} W^s(x, f, g).$$

We want to rephrase the definition of unstable and stable manifolds in terms of spaces of negative half-trajectories as it is done in [Sch99]. This reformulation will be helpful for introducing the notion of perturbed gradient flow half-trajectories. Define

$$\overline{\mathbb{R}} := \mathbb{R} \cup \{-\infty, +\infty\}, \quad \overline{\mathbb{R}}_{\geq 0} := [0, +\infty) \cup \{+\infty\}, \quad \overline{\mathbb{R}}_{\leq 0} := (-\infty, 0] \cup \{-\infty\}.$$

Following [Sch93, Sect. 2.1], we equip $\overline{\mathbb{R}}$ with a smooth structure by choosing a diffeomorphism $h : \mathbb{R} \to (-1, 1)$ with

$$h(0) = 0, \quad \lim_{s \to -\infty} h(s) = -1, \quad \lim_{s \to +\infty} h(s) = 1,$$

by extending h in the obvious way to a bijection $h : \overline{\mathbb{R}} \to [-1, 1]$ and requiring that this extended map is a diffeomorphism of manifolds with boundary. In the following we always assume such a map h to be chosen and fixed. We equip $\overline{\mathbb{R}}_{\leq 0}$ and $\overline{\mathbb{R}}_{\geq 0}$ with the unique smooth structures which make them smooth submanifolds with boundary of $\overline{\mathbb{R}}$.

For a Riemannian metric g on M and $x \in \text{Crit } f$ define spaces of curves

$$\mathcal{P}_-(x) := \left\{ \gamma \in H^1 \left(\overline{\mathbb{R}}_{\leq 0}, M \right) \Big| \lim_{s \to -\infty} \gamma(s) = x \right\},$$

$$\mathcal{P}_+(x) := \left\{ \gamma \in H^1 \left(\overline{\mathbb{R}}_{\geq 0}, M \right) \Big| \lim_{s \to +\infty} \gamma(s) = x \right\},$$

where H^1 always denotes spaces of maps of Sobolev class $W^{1,2}$ with respect to the Riemannian metric g. These spaces can be equipped with structures of Hilbert manifolds, see [Sch93, Appendix A]. The following spaces can be equipped with structures of Banach bundles over $\mathcal{P}_-(x)$, $\mathcal{P}_+(x)$, respectively, by

$$\mathcal{L}_-(x) := \bigcup_{\gamma \in \mathcal{P}_-(x)} L^2(\gamma^* TM), \quad \mathcal{L}_+(x) := \bigcup_{\gamma \in \mathcal{P}_+(x)} L^2(\gamma^* TM),$$

together with the obvious bundle projections, where $L^2(\gamma^* TM)$ denotes the space of sections of the vector bundle $\gamma^* TM$ which are L^2-integrable with respect to g. It can be shown by the methods of [Sch93, Appendix A] that $\mathcal{L}_-(x) \to \mathcal{P}_-(x)$ and $\mathcal{L}_+(x) \to \mathcal{P}_+(x)$ are indeed smooth Banach bundles over Hilbert manifolds.

The following notion is implicitly used in [Sch93, Sch99], though its name seems to have its first appearance in the literature in [HWZ07] in a much more general context. We will use it in our Morse-theoretic framework in the upcoming theorems.

Definition 2.2 Let M be a Banach manifold and $E \to M$ be a Banach bundle over M with typical fiber F. A section of the bundle $s : M \to E$ is called a *Fredholm section of index* $r \in \mathbb{Z}$, if there is a local trivialization

$$\left\{ \left(U_\alpha, \varphi_\alpha : E|_{U_\alpha} \overset{\cong}{\to} U_\alpha \times F \right) \right\}_{\alpha \in I},$$

where I is an index set, such that for every $\alpha \in I$ the map

$$pr_2 \circ \varphi_\alpha \circ s|_{U_\alpha} : U_\alpha \to F,$$

where $pr_2 : U_\alpha \times F \to F$ denotes the projection onto the second factor, is a Fredholm map of index r.

The following theorem is well-known in Morse theory and can be proven by the methods of [Sch93, Sect. 2.2]:

Theorem 2.3   *1. The map*

$$\partial^g : \mathcal{P}_-(x) \to \mathcal{L}_-(x) , \quad \gamma \mapsto \dot{\gamma} + \nabla^g f \circ \gamma ,$$

is a smooth section of $\mathcal{L}_-(x)$ and a Fredholm section of index

$$\text{ind } \partial^g = \mu(x) .$$

Furthermore, the space

$$\mathcal{W}^u(x, f, g) := \{\gamma \in \mathcal{P}_-(x) \mid \dot{\gamma} + \nabla^g f \circ \gamma = 0\} = (\partial^g)^{-1}(\mathbb{O}_{\mathcal{L}_-(x)})$$

is a submanifold of $\mathcal{P}_-(x)$ with dim $\mathcal{W}^u(x, f, g) = \mu(x)$, where $\mathbb{O}_{\mathcal{L}_-(x)}$ denotes the image of the zero section of $\mathcal{L}_-(x)$.
2. *The map*
$$\partial^g : \mathcal{P}_+(x) \to \mathcal{L}_+(x) , \quad \gamma \mapsto \dot{\gamma} + \nabla^g f \circ \gamma ,$$

is a smooth section of $\mathcal{L}_+(x)$ and a Fredholm section of index

$$\text{ind } \partial^g = n - \mu(x) .$$

Furthermore, the space

$$\mathcal{W}^s(x, f, g) := \{\gamma \in \mathcal{P}_+(x) \mid \dot{\gamma} + \nabla^g f \circ \gamma = 0\} = (\partial^g)^{-1}(\mathbb{O}_{\mathcal{L}_+(x)})$$

is a submanifold of $\mathcal{P}_+(x)$ with dim $\mathcal{W}^s(x, f, g) = n - \mu(x)$, where $\mathbb{O}_{\mathcal{L}_+(x)}$ denotes the image of the zero section of $\mathcal{L}_+(x)$.

Remark 2.4   1. It is well-known that the evaluation map

$$\mathcal{W}^u(x, f, g) \to M , \quad \gamma \mapsto \gamma(0) ,$$

is an embedding onto $W^u(x, f, g)$ and that the map $\mathcal{W}^s(x, f, g) \to M, \gamma \mapsto \gamma(0)$, is an embedding onto $W^s(x, f, g)$.
2. Using standard regularity results for solutions of ordinary differential equations, one can show that $\mathcal{W}^u(x, f, g)$ and $\mathcal{W}^s(x, f, g)$ consist of smooth curves only, see [Sch93, Proposition 2.9]. More precisely, we can identify as sets:

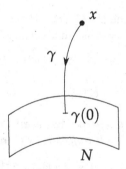

Fig. 2.1 An element of $\{\gamma \in \mathcal{W}^u(x, f, g) \mid \gamma(0) \in N\}$

$$\mathcal{W}^u(x, f, g) = \left\{\gamma \in C^\infty\left((-\infty, 0], M\right) \;\middle|\; \lim_{s \to -\infty} \gamma(s) = x, \; \dot{\gamma} + \nabla^g f \circ \gamma = 0\right\},$$

$$\mathcal{W}^s(x, f, g) = \left\{\gamma \in C^\infty\left([0, +\infty), M\right) \;\middle|\; \lim_{s \to +\infty} \gamma(s) = x, \; \dot{\gamma} + \nabla^g f \circ \gamma = 0\right\}.$$

In the course of this chapter, we will consider moduli spaces which are similar to those of the form

$$\{\gamma \in \mathcal{W}^u(x, f, g) \mid \gamma(0) \in N\},$$

where N is an arbitrary smooth submanifold of M, see Fig. 2.1. More precisely, we would like to use transversality theorems to show that for a generic choice of Riemannian metric g, the former space is a smooth manifold.

A problem that occurs is that we will not be able to apply transversality theory and in particular the Sard–Smale transversality theorem if N contains critical points of f without an additional condition on the Morse index of x. (For statements in this direction see [Sch99, Lemma 4.10] and [AS10, Appendix A.2].)

2.2 Perturbed Negative Semi-infinite Trajectories

We overcome this transversality problem by using time-dependent vector fields to perturb the negative gradient flow equation. This method was introduced by Mohammed Abouzaid in [Abo11]. Since Abouzaid only sketches his construction, we will carry out the setting in greater detail.

Define

$$\mathfrak{X}_-(M) := \left\{X \in C^{n+1}\left((-\infty, 0] \times M, TM\right) \;\middle|\; X(s, \cdot) \in C^{n+1}(TM) \;\; \forall s \in (-\infty, 0],\right.$$

$$\left. X(s, x) = 0 \;\; \forall s \leq -1, \; x \in M\right\},$$

where $C^{n+1}(TM)$ denotes the space of sections of the bundle TM of class C^{n+1}, i.e. the space of vector fields of class C^{n+1}. Then $\mathfrak{X}_-(M)$ is a closed linear subspace of the Banach space of C^{n+1}-sections of $\pi^* TM$, where $\pi : (-\infty, 0] \times M \to M$ denotes the projection and where we equip $C^{n+1}(\pi^* TM)$ with the C^{n+1}-norm induced by the chosen metric g on M and the standard metric on $(-\infty, 0]$.

Definition 2.5 Let $x \in \mathrm{Crit}\, f$ and $X \in \mathfrak{X}_-(M)$. A curve $\gamma \in \mathcal{P}_-(x)$ which satisfies the equation

$$\dot{\gamma}(s) + \nabla^g f(\gamma(s)) + X(s, \gamma(s)) = 0$$

for every $s \in (-\infty, 0]$ is called a *perturbed negative gradient flow half-trajectory* (or simply a *perturbed negative half-trajectory*) with respect to (f, g, X).

Remark 2.6 We consider vector fields of class C^{n+1} as perturbations instead of smooth vector fields, since we want to momentarily apply the Sard–Smale transversality theorem. This theorem requires the space of parameters to be a Banach manifold modelled on a separable Banach space. While this property obviously holds true for $\mathfrak{X}_-(M)$, the subspace of smooth vector fields in $\mathfrak{X}_-(M)$ fails to be complete, as it is well-known from functional analysis.

There are two ways to overcome this problem in the literature, see [Flo88c] and [Sch93, Sect. 2.3] for the one approach and [FHS95] for the other. Since for every important result in this book it suffices to consider time-dependent vector fields which are finitely many times differentiable, we will not further elaborate on these approaches.

Moreover, we choose the perturbations to be $(n + 1)$ times differentiable, since this is the minimal amount of regularity that will be required in our situation for the application of the Sard–Smale transversality theorem in the proof of Theorem 2.14.

Throughout the rest of this chapter, we fix a Riemannian metric g on M and will mostly leave it out of the notation.

In the following, we write $\mathfrak{X}_- := \mathfrak{X}_-(M)$.

Theorem 2.7 *Let $x \in \mathrm{Crit}\, f$. The map*

$$F^- : \mathfrak{X}_- \times \mathcal{P}_-(x) \to \mathcal{L}_-(x)\,, \quad (Y, \gamma) \mapsto (s \mapsto \dot{\gamma}(s) + \nabla f(\gamma(s)) + Y(s, \gamma(s)))\,,$$

is an $(n + 1)$ times differentiable map of between Banach manifolds and for every $Y \in \mathfrak{X}_-$, the map $F_Y^- := F^-(Y, \cdot)$ is a C^{n+1}-section of $\mathcal{L}_-(x)$. Furthermore, every F_Y^- is a Fredholm section of index

$$\mathrm{ind}\, F_Y^- = \mu(x)\,.$$

Remark 2.8 Note that if $0 \in \mathfrak{X}_-$ denotes the zero vector field, then

$$F_0^- = \partial^g\,. \tag{2.1}$$

Thus, Theorem 2.7 generalizes Theorem 2.3.

Proof of Theorem 2.7 For a fixed $Y \in \mathfrak{X}_-$, one shows in in strict analogy with the proof of [Sch93, Theorem 12] that the map $F^-(Y, \cdot) : \mathcal{P}_-(x) \to \mathcal{L}_-(x)$ is $(n + 1)$

times differentiable for every $Y \in \mathfrak{X}_-$. Moreover, F^- is continuous and affine, hence smooth, in the $\mathfrak{X}_-$-component. So F^- is $(n+1)$ times differentiable by the total differential theorem.

The Fredholm property of the F_Y^- follows in precisely the same way as in the unperturbed case in part 1 of Theorem 2.3, i.e. by similar methods as in the proof of [Sch93, Proposition 2.2].

By part 1 of Theorem 2.3, the operator $F_0^- = \partial^g$ has index $\mu(x)$. Since $\mathfrak{X}_-$ is connected and the Fredholm index is locally constant, it follows that ind $F_Y^- = \mu(x)$ for every $Y \in \mathfrak{X}_-$. $\qquad\square$

Theorem 2.9 *The map $F_Y^- : \mathcal{P}_-(x) \to \mathcal{L}_-(x)$ is transverse to $\mathcal{O}_{\mathcal{L}_-(x)}$ for all $Y \in \mathfrak{X}_-$ and $x \in \mathrm{Crit}\, f$.*

Proof This can be shown by transferring the methods used in [Jos08, Sect. 6.3] and in [AM06, Sect. 2] to the formalism of [Sch93]. Compared to [Jos08, Sect. 6.3], we need to replace the flow of the vector field $-\nabla f$ by the time-dependent flow of the time-dependent vector field $-\nabla f - X$ and repeat the line of arguments used to prove [Jos08, Corollary 6.3.1] (see the preparations of Proposition 2.19 below for a more detailed discussion of flows of time-dependent vector fields). $\qquad\square$

Corollary 2.10 *For every $Y \in \mathfrak{X}_-$, the space*

$$W^-(x, Y) := \{\gamma \in \mathcal{P}_-(x) \mid \dot{\gamma}(s) + \nabla f(\gamma(s)) + Y(s, \gamma(s)) = 0 \ \forall s \in (-\infty, 0]\}$$

is a submanifold of $\mathcal{P}_-(x)$ of class C^{n+1} with $\dim W^-(x, Y) = \mu(x)$.

Proof This follows from Theorem 2.9 and the observation that

$$W^-(x, Y) = (F_Y^-)^{-1}(\mathcal{O}_{\mathcal{L}_-(x)}) .$$

$\qquad\square$

Remark 2.11 This corollary generalizes the submanifold statement of part 1 of Theorem 2.3, since it obviously holds that

$$W^-(x, 0) = \mathcal{W}^u(x, f, g) .$$

So far, we have shown that the space of perturbed negative half-trajectories emanating from a given critical point can be described as a differentiable manifold. The introduction of time-dependent perturbations gives us a much higher flexibility considering evaluation maps and couplings with other trajectory spaces. We will see this in the next theorems and in full generality in Chap. 3.

Remark 2.12 The following differences occur between the space of unperturbed negative half-trajectories $\mathcal{W}^u(x, f, g)$ and the perturbed trajectory spaces $W^-(x, Y)$:

1. While the constant half-trajectory $(t \mapsto x)$ is always an element of $\mathcal{W}^u(x, f, g)$, it is not an element of $W^-(x, Y)$ for $Y \in \mathfrak{X}_-$ if there is some $s \in (-1, 0]$ such that $Y(s, x) \neq 0$.

2. While $\mathcal{W}^u(x, f, g)$ will be a smooth submanifold of $\mathcal{P}_-(x)$ if the Morse function f is smooth, the differentiability of $W^-(x, Y)$ is always given by the differentiability of Y. For our later considerations, it will only be of importance for $W^-(x, Y)$ to be of class C^1. This is guaranteed for our choices of Y.

Define

$$\widetilde{W}^-(x, \mathfrak{X}_-) := \left(F^-\right)^{-1}\left(\mathcal{O}_{\mathcal{L}_-(x)}\right) \subset \mathfrak{X}_- \times \mathcal{P}_-(x) .$$

Since F_Y^- is transverse to $\mathcal{O}_{\mathcal{L}_-(x)}$ for every $Y \in U$, the total map F^- is transverse to $\mathcal{O}_{\mathcal{L}_-(x)}$, so $\widetilde{W}^-(x, \mathfrak{X}_-)$ is a Banach submanifold of $\mathfrak{X}_- \times \mathcal{P}_-(x)$.

The following statement is a straightforward analogue of [Sch99, Lemma 4.10] for perturbed half-trajectories and we will prove it by the same methods.

Theorem 2.13 *The map*

$$E^- : \widetilde{W}^-(x, \mathfrak{X}_-) \to M , \quad (Y, \gamma) \mapsto \gamma(0) ,$$

is a submersion of class C^{n+1}.

Proof We start by determining the tangent bundle of $\widetilde{W}^-(x, \mathfrak{X}_-)$. Let ∇ be a Riemannian connection on TM. (To distinguish the notation from ∇f, we will write $\nabla^g f$ for the gradient vector field with respect to g in this proof.)

One uses the methods of [Sch93, Appendix A] and the proof of [Sch99, Lemma 4.10] as well as the simple fact that $F^-(\cdot, \gamma)$ is an affine map for every $\gamma \in \mathcal{P}_-(x)$ to show that the differential of F^- is given by

$$\left(DF^-_{(Y,\gamma)}(Z, \xi)\right)(t) = (\nabla_t \xi)(t) + \left(\nabla_\xi \nabla^g f\right)(\gamma(t)) + \left(\nabla_\xi Y\right)(\gamma(t)) + Z(t, \gamma(t)) , \tag{2.2}$$

at every $(Y, \gamma) \in U$ and for all $Z \in T_Y \mathfrak{X}_- \cong \mathfrak{X}_-, \xi \in T_\gamma \mathcal{P}_-(x)$, where $\nabla_t \xi$ denotes the covariant derivative of ξ along the curve γ. By definition of $\widetilde{W}^-(x, \mathfrak{X}_-)$ as a submanifold of the product, we know that

$$T_{(Y,\gamma)} \widetilde{W}^-(x, \mathfrak{X}_-) = \ker(DF^-)_{(Y,\gamma)}$$
$$= \left\{(Z, \xi) \in T_{(Y,\gamma)}(\mathfrak{X}_- \times \mathcal{P}_-(x)) \mid \nabla_t \xi + \left(\nabla_\xi \nabla^g f\right) \circ \gamma + \left(\nabla_\xi Y\right) \cdot \circ \gamma + Z \circ \gamma = 0\right\} .$$

Moreover, we can extend the map E^- in the obvious way to a map defined on the whole product $E^- : \mathfrak{X}_- \times \mathcal{P}_-(x) \to M$ and for all $Z \in T_Y \mathfrak{X}_-$ and $\xi \in T_\gamma \mathcal{P}_-(x)$, the differential of E^- is given by

$$DE^-_{(Y,\gamma)}[(Z, \xi)] = \xi(0) . \tag{2.3}$$

Let $v \in T_{\gamma(0)} M$ be given and choose $\xi_0 \in T_\gamma \mathcal{P}_-(x)$ with the following properties:

$$\xi_0 \in C^{n+1}(\gamma^* TM), \tag{2.4}$$

$$\left(\nabla_t \xi_0 + \left(\nabla_{\xi_0} \nabla^g f\right) \circ \gamma + \left(\nabla_{\xi_0} Y\right) \circ \gamma\right)(s) = 0 \quad \forall s \in (-\infty, -1], \tag{2.5}$$

$$\xi_0(0) = v. \tag{2.6}$$

The existence of a $\xi_0 \in T_\gamma \mathcal{P}_-(x)$ satisfying (2.4), (2.5) and (2.6) is easy to see: We define ξ_0 on $(-\infty, -1]$ as an arbitrary solution of the first order ordinary differential equation (2.5) which exists by the standard results for uniqueness and existence of solutions. One can then show iteratively (similar as in the proof of [Sch93, Proposition 2.9]), using that f is smooth and Y is of class C^{n+1}, that ξ_0 can be chosen to be $(n+1)$ times differentiable on $(-\infty, -1]$. Moreover, one applies [Sch93, Lemma 2.10] to solutions of (2.5) and derives that $\|\xi_0\|_{C^{n+1}} < +\infty$. This implies (2.4).

We continue this ξ_0 by an arbitrary vector field along $\gamma|_{[-1,0]}$ such that (2.4) and (2.6) are satisfied. By (2.3) and (2.6), we know that

$$DE^-_{(Y,\gamma)}[(0, \xi_0)] = v.$$

We then pick $Z_0 \in \mathcal{X}_-$ with the following property:

$$Z_0(t, \gamma(t)) = -(DF^-_{(Y,\gamma)}[(0, \xi_0)])(t) \quad \forall t \in (-\infty, 0].$$

This property does not contradict the definition of $\mathcal{X}_-$ since (2.5) is equivalent to

$$(DF^-_{(Y,\gamma)}[(0, \xi_0)])(t) = 0 \quad \text{for every } t \in (-\infty, -1].$$

The existence of such a Z_0 is obvious. By the definitions of ξ_0 and Z_0 we obtain using (2.2) and (2.3):

$$DE^-_{(Y,\gamma)}[(Z_0, \xi_0)] = v, \qquad DF^-_{(Y,\gamma)}[(Z_0, \xi_0)] = 0.$$

The last line implies that $(Z_0, \xi_0) \in T_{(Y,\gamma)} \widetilde{W}^-(x, U)$. So for an arbitrary $v \in T_{\gamma(0)}M$ we have found such a (Z_0, ξ_0) which maps to v under DE^- and therefore proven the claim. □

The following theorem is derived from Theorem 2.13 by standard methods of proving transversality results in Morse and Floer homology (see e.g. [Sch93, Sect. 2.3], [FHS95] or [Hut02, Sect. 5]).

Theorem 2.14 *Let $N \subset M$ be a closed submanifold. There is a generic set $\mathcal{G} \subset \mathcal{X}_-$ such that for every $Y \in \mathcal{G}$, the map*

$$E^-_Y : W^-(x, Y) \to M, \quad \gamma \mapsto E^-(Y, \gamma),$$

is transverse to N. Consequently, for $Y \in \mathcal{G}$, the space

$$W^-(x, Y, N) := \{\gamma \in W^-(x, Y) \mid \gamma(0) \in N\} = \left(E_Y^-\right)^{-1}(N)$$

is a manifold of class C^{n+1} with

$$\dim W^-(x, Y, N) = \mu(x) - \operatorname{codim}_M N .$$

Proof We want to apply the Sard–Smale transversality theorem in the form of [Sch93, Proposition 2.24] or [Hut02, Theorem 5.4] using Theorem 2.13. For this purpose, we have to express $W^-(x, Y, N)$ in a slightly different way. Define

$$\mathbb{O}_{\mathcal{L}_-(x),N} := \{(\gamma, 0) \in \mathcal{L}_-(x) \mid \gamma(0) \in N\} .$$

Since the map $\mathcal{P}_-(x) \to M, \gamma \mapsto \gamma(0)$, is a submersion, $\mathbb{O}_{\mathcal{L}_-(x),N}$ is the image of the restriction of the zero-section to a Banach submanifold of $\mathcal{P}_-(x)$ whose codimension is given by codim N.

Comparing the definitions of both sides of the following equation, one checks that for every $Y \in \mathfrak{X}_-$ we have

$$E_Y^{-1}(N) = \left(F_Y^-\right)^{-1}(\mathbb{O}_{\mathcal{L}_-(x),N}) . \tag{2.7}$$

But it is easy to see that Theorems 2.9 and 2.13 together imply that the total map F^- is transverse to $\mathbb{O}_{\mathcal{L}_-(x),N}$. The map F^- is defined on the product $\mathfrak{X}_- \times \mathcal{P}_-(x)$. Moreover, F^- is a Fredholm map of index

$$0 \le \mu(x) \le n$$

by Theorem 2.7. The same theorem implies that F^- is of class C^{n+1} which shows that the differentiability requirements of the Sard–Smale transversality theorem are satisfied by F^-. Viewing $\mathfrak{X}_-$ as a parameter space, one sees that Theorem 2.13 implies all the remaining requirements for the Sard–Smale theorem. We derive:

There is a generic subset $\mathcal{G}_x \subset \mathfrak{X}_-$ such that for $Y \in \mathcal{G}_x$, the map F_Y^- is transverse to $\mathbb{O}_{\mathcal{L}_-(x),N}$, so it follows that for $Y \in \mathcal{G}_x$, the space $\left(E_Y^-\right)^{-1}(N) = W^-(x, Y, N)$ is a manifold of class C^{n+1} with

$$\dim W^-(x, Y, N) = \operatorname{ind} F_Y^- - \operatorname{codim}_{\mathbb{O}_{\mathcal{L}_-(x)}} \mathbb{O}_{\mathcal{L}_-(x),N}$$
$$= \mu(x) - \operatorname{codim}_M N .$$

We have so far constructed an individual generic set $\mathcal{G}_x$ for each $x \in \operatorname{Crit} f$. Define

$$\mathcal{G} := \bigcap_{x \in \operatorname{Crit} f} \mathcal{G}_x \subset \mathfrak{X}_-(M) .$$

Since Crit f is a finite set, $\mathcal{G}$ is a finite intersection of generic sets, hence itself generic. Moreover, for every $Y \in \mathcal{G}$, the map E_Y^- is transverse to N. $\square$

2.3 Perturbed Positive Semi-infinite Trajectories

We obtain analogous results for positive half-trajectories, i.e. for analogous curves of the form

$$[0, +\infty) \to M .$$

We next give a brief account of these results and omit the proofs, since they can be done along the same lines as for perturbed negative half-trajectories.

The analogous space of perturbations for curves $[0, \infty) \to M$ is given by:

$$\mathfrak{X}_+(M) := \left\{ Y \in C^{n+1}([0, +\infty) \times M, TM) \,\middle|\, Y(s, \cdot) \in C^{n+1}(TM) \quad \forall s \in [0, +\infty), \right.$$
$$\left. Y(s, x) = 0 \quad \forall s \geq 1, \ x \in M \right\} .$$

We summarize the corresponding results in the following theorem. It can be derived from part 2 of Theorem 2.3 in the same way as Corollary 2.10 and Theorems 2.13 and 2.14 are derived from part 1 of Theorem 2.3.

Theorem 2.15 *1. For all $Y \in \mathfrak{X}_+(M)$ and $x \in \operatorname{Crit} f$ the space*

$$W^+(x, Y) := \{\gamma \in \mathcal{P}_+(x) \mid \dot{\gamma}(s) + \nabla f(\gamma(s)) + Y(s, \gamma(s)) = 0\}$$

is a submanifold of $\mathcal{P}_+(x)$ of class C^{n+1} with

$$\dim W^+(x, Y) = n - \mu(x) .$$

2. For $x \in \operatorname{Crit} f$ define $\widetilde{W}^+(x, \mathfrak{X}_+) := \{(Y, \gamma) \in \mathfrak{X}_+(M) \times \mathcal{P}_+(x) \mid \gamma \in W^+(x, Y)\}$. The space $\widetilde{W}^+(x, \mathfrak{X}_+)$ is a Banach submanifold of $\mathfrak{X}_+(M) \times \mathcal{P}_+(x)$, and the map

$$E^+ : \widetilde{W}^+(x, \mathfrak{X}_+) \to M , \quad (Y, \gamma) \mapsto \gamma(0) ,$$

is a submersion of class C^{n+1}.

3. Let $N \subset M$ be a closed submanifold. There is a generic subset $\mathcal{G} \subset \mathfrak{X}_+$ such that for all $Y \in \mathcal{G}$ and $x \in \operatorname{Crit} f$ the space

$$W^+(x, Y, N) := \{\gamma \in W^+(x, Y) \mid \gamma(0) = x\}$$

is a manifold of class C^{n+1} of dimension

$$\dim W^+(x, Y, N) = \dim N - \mu(x) .$$

2.4   Perturbed Finite-Length Trajectories

We consider a third type of perturbed trajectories of the negative gradient flow of f, namely curves $\gamma : [0, l] \to M$, where $l \in [0, +\infty)$ is allowed to vary, satisfying a perturbed negative gradient flow equation. Once again we start by considering the unperturbed case and derive our results for perturbed trajectories from this case.

Consider the set

$$\mathcal{M}(f, g) := \left\{ (l, \gamma) \mid l \in [0, +\infty) , \; \gamma : [0, l] \to M , \; \dot{\gamma} + \nabla^g f \circ \gamma = 0 \right\} .$$

By the unique existence of solutions of ordinary differential equations for given initial values, the following map is easily identified as a bijection:

$$\varphi : \mathcal{M}(f, g) \to [0, +\infty) \times M , \quad (l, \gamma) \mapsto (l, \gamma(0)) , \tag{2.8}$$

whose inverse is given by the map

$$\varphi^{-1} : [0, +\infty) \times M \to \mathcal{M}(f, g) , \quad (l, x) \mapsto \left(l, \left(t \mapsto \phi_t^{-\nabla^g f}(x), \; t \in [0, l] \right) \right) ,$$

where $\phi^{-\nabla^g f}$ again denotes the negative gradient flow of f with respect to g.

We equip $[0, +\infty)$ with the canonical structure of a smooth manifold with boundary. Then the product manifold $[0, +\infty) \times M$ is a smooth manifold with boundary as well.

We can therefore equip $\mathcal{M}(f, g)$ with a topology and a smooth structure such that φ becomes a diffeomorphism of manifolds with boundary. From now on, let $\mathcal{M}(f, g)$ always be equipped with this smooth structure.

Definition 2.16 We call $\mathcal{M}(f, g)$ the space of *finite-length trajectories* of the negative gradient flow of f.

Spaces of finite-length trajectories as well as their smooth structures are also considered in [Weh12, Sect. 2] in greater detail. To study perturbed finite-length gradient flow trajectories, we momentarily introduce a convenient space of perturbations. The construction of the perturbation spaces requires more effort since we want the interval length of the trajectory to be a parameter of the perturbing vector fields as well. This becomes necessary in Chap. 5 where we consider the limiting behaviour of sequences of perturbed finite-length trajectories if the length tends to $+\infty$.

Consider a smooth function $\chi : [0, +\infty) \to [0, 1]$ with the following properties:

- $\chi(l) = \frac{1}{3}$ if $l \leq 3 - \delta$ for some small $\delta > 0$,
- $\chi(l) = 1$ if $l \geq 3$,
- $\dot{\chi}(l) > 0$ if $l < 3$.

See Fig. 2.2 for a picture the graph of a possible choice of χ.

We choose such a function χ once and for all and keep it fixed throughout the rest of this book.

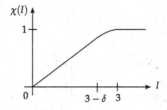

Fig. 2.2 The graph of χ

Moreover, for any map

$$Z : [0, +\infty) \times [0, +\infty) \times M \to TM , \quad (l, t, x) \mapsto Z(l, t, x) ,$$

and any $l \geq 1$ we define maps

$$s_0(l, Z) : [0, +\infty) \times M \to TM , \quad (t, x) \mapsto \begin{cases} Z(l, t, x) & \text{if } t \in [0, 1] , \\ 0 \in T_x M & \text{if } t > 1 , \end{cases}$$

$$e_0(l, Z) : (-\infty, 0] \times M \to TM , \quad (t, x) \mapsto \begin{cases} Z(l, t+l, x) & \text{if } t \in [-1, 0] , \\ 0 \in T_x M & \text{if } t < -1 . \end{cases}$$

One checks from these definitions that if Z is chosen such that $Z(l, t, \cdot)$ is a vector field on M for all $l, t \in [0, +\infty)$, then both $s_0(l, Z)$ and $e_0(l, Z)$ are time-dependent sections of TM, although they might not be continuous.

We further put for all $l \in [0, +\infty)$ and $Z : [0, +\infty) \times [0, +\infty) \times M \to TM$ with the above properties:

$$\text{split}_0(l, Z) := (s_0(l, Z), e_0(l, Z)) . \tag{2.9}$$

Define

$$\mathfrak{X}_0(M) := \Big\{ X \in C^{n+1}\left([0, +\infty)^2 \times M, TM\right) \,\Big|\, X(l, t, \cdot) \in C^{n+1}(TM) \,\forall\, l,$$

$$t \in [0, +\infty) , X(l, t, x) = 0 \text{ if } t \in [\chi(l), l - \chi(l)] , \tag{2.10}$$

$$(D^k X)_{(0,t,x)} = 0 \ \forall k \in \{0, 1, \ldots, n+1\}, \ (t, x) \in [0, +\infty) \times M , \tag{2.11}$$

$$\lim_{l \to +\infty} \text{split}_0(l, Y) \text{ exists in } \mathfrak{X}_+(M) \times \mathfrak{X}_-(M) \Big\} . \tag{2.12}$$

Concerning condition (2.10), it suffices for the rest of this section to note that this condition especially implies that for each $l \in \mathbb{R}_{>0}$ there is a $\delta > 0$ such that

$$X(l, t, x) = 0 \quad \text{for all} \quad t \in \left(\tfrac{l}{2} - \delta, \tfrac{l}{2} + \delta\right), \quad x \in M \text{ and } X \in \mathfrak{X}_0(M) .$$

Moreover, if $l \geq 3$, then

$$X(l, t, x) = 0 \quad \text{for every} \quad t \in [1, l - 1] . \tag{2.13}$$

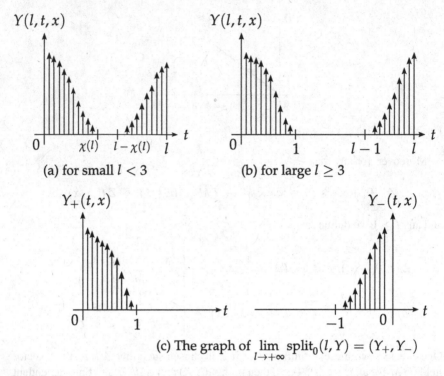

Fig. 2.3 The values of $t \mapsto Y(l, t, x)$ for $Y \in \mathfrak{X}_0(M)$ and fixed $l \geq 0$, $x \in M$ and an illustration of the map split_0

See Fig. 2.3 for an illustration of these observations. Condition (2.11) will not be required until Chap. 6, see Remark 6.8. From (2.13) and the definition of the maps s_0 and e_0 we derive the following observation.

Lemma 2.17 $\mathfrak{X}_0(M)$ *is a Banach space with the* C^{n+1}*-norm induced by the given metrics.*

Proof Apparently, $\mathfrak{X}_0(M)$ is a subset of $C^{n+1}(\pi^*TM)$, where $\pi : [0, +\infty)^2 \times M \to M$ denotes the projection onto the second factor. One checks without difficulties that the space

$$\tilde{\mathfrak{X}}_0(M) := \left\{ X \in C^{n+1}(\pi^*TM) \,\middle|\, X \text{ satisfies (2.10) and (2.11)} \right\}$$

is a closed linear subspace of $C^{n+1}(\pi^*TM)$. One shows that for $l \geq 3$ and $X \in \tilde{\mathfrak{X}}_0(M)$, the vector fields $s_0(l, X)$ and $e_0(l, x)$ from above are of class C^{n+1}, implying that the map

$$\mathfrak{X}_0(M) \to \mathfrak{X}_+(M) \times \mathfrak{X}_-(M), \quad X \mapsto \mathrm{split}_0(l, X) \,,$$

is well-defined and continuously linear for every $l \geq 3$. It follows from this observation and the definition of $\mathfrak{X}_0(M)$ that the latter is a closed linear subspace of $\tilde{\mathfrak{X}}_0(M)$, hence of $C^{n+1}(\pi^* TM)$, inheriting a Banach space structure. $\square$

Remark 2.18   One derives from (2.13) that for any $Y \in C^{n+1}([0, +\infty) \times [0, +\infty) \times M, TM)$ which obeys condition (2.10) and for which the limit $Y_0 := \lim_{l \to +\infty} \text{split}_0(l, Y)$ exists pointwise, the pair of time-dependent vector fields Y_0 is of class C^{n+1} and lies in $\mathfrak{X}_+(M) \times \mathfrak{X}_-(M)$. This observation is the reason for introducing condition (2.10), since it will be needed for the compactifications of certain moduli spaces in the upcoming sections.

For any l we can view X_l as a time-dependent vector field on M, where

$$X_l(t, x) := X(l, t, x) \quad \forall t \in [0, +\infty), \quad x \in M .$$

To proceed, we briefly recall the notion of a flow of a time-dependent vector field. For details see [Lee03, Chap. 17]. For a time-dependent vector field $Y : \mathbb{R} \times M \to TM$, an *integral curve of Y* is defined to be a curve $\alpha : I \to M, I \subset \mathbb{R}$ for which

$$\dot{\alpha}(t) = Y(t, \alpha(t))$$

holds for every $t \in I$. An elementary result from the theory of flows implies that every integral curve of a time-dependent vector field on a complete manifold can be extended to an integral curve defined on the whole real line.

The *time-dependent flow of Y* is the map $\phi^Y : \mathbb{R} \times \mathbb{R} \times M \to M$ that is uniquely defined by the following two properties:

- $\phi_{t,t}^Y(x) = x$ for every $t \in \mathbb{R}$ and $x \in M$,
- For every $t \in \mathbb{R}$ and $x \in M$ the map $\mathbb{R} \to M, s \mapsto \phi_{s,t}^Y(x)$ is the unique solution of

$$\begin{cases} \gamma(t) = x , \\ \dot{\gamma}(s) = Y(s, \gamma(s)) \quad \forall s \in \mathbb{R} , \end{cases}$$

where $\phi_{s,t}^Y(x) := \phi^Y(s, t, x)$ for every $s, t \in [0, +\infty)$ and $x \in M$. Consider the space

$$\widetilde{\mathcal{M}} := \{(Y, l, \gamma) \mid Y \in \mathfrak{X}_0(M),\ l \in [0, +\infty),\ \gamma : [0, l] \to M,$$
$$\dot{\gamma}(s) + (\nabla f)(\gamma(s)) + Y_l(s, \gamma(s)) = 0\} . \tag{2.14}$$

The following map is then easily identified as a bijection:

$$\psi : \mathfrak{X}_0(M) \times \mathcal{M}(f, g) \to \widetilde{\mathcal{M}} ,$$
$$(Y, l, \gamma) \mapsto \left(Y, \left(l, \left(s \mapsto \phi_{s,\frac{l}{2}}^{Y,l}\left(\gamma\left(\tfrac{l}{2}\right) \right) \right),\ s \in [0, l] \right) \right) , \tag{2.15}$$

where $\phi_{s,t}^{Y,l}$ denotes the flow of the time-dependent vector field

$$(s, x) \mapsto -Y_l(s, x) - (\nabla f)(x) \,.$$

We want to describe ψ more intuitively. First of all, for any $x \in M$, by definition of the flow the curve $\alpha_x : [0, l] \to M$,

$$\alpha_x(s) := \phi_{s, \frac{l}{2}}^{Y, l}(x) \,,$$

denotes the unique solution of (2.14) which is defined on $[0, l]$ and which fulfills

$$\alpha_x \left(\frac{l}{2} \right) = x \,.$$

Therefore, if $(Y, l, \gamma) \in \mathfrak{X}_0(M) \times \mathcal{M}(f, g)$ and if we define $\tilde{\gamma} : [0, l] \to M$ by putting

$$(Y, l, \tilde{\gamma}) := \psi(Y, l, \gamma) \,,$$

then $\tilde{\gamma}$ will be the unique solution of (2.14) with

$$\tilde{\gamma} \left(\frac{l}{2} \right) = \gamma \left(\frac{l}{2} \right) \,.$$

Loosely speaking, ψ maps a solution of the negative gradient flow equation to the solution of (2.14) defined on the same interval and coinciding with γ at time $\frac{l}{2}$.

Moreover, the inverse of ψ is given by the map $\widetilde{\mathcal{M}} \to \mathfrak{X}_0(M) \times \mathcal{M}(f, g)$,

$$(Y, (l, \gamma)) \to \left(Y, \left(l, \left(s \mapsto \phi_{s - \frac{l}{2}}^{-\nabla f} \left(\gamma \left(\tfrac{l}{2} \right) \right) \right), s \in [0, l] \right) \right) \,.$$

Since ψ is a bijection and its domain is the product of a Banach space and a smooth Banach manifold with boundary, we can equip $\widetilde{\mathcal{M}}$ with the unique structure of a Banach manifold with boundary such that ψ becomes a smooth diffeomorphism of Banach manifolds with boundary. Let $\widetilde{\mathcal{M}}$ always be equipped with this smooth structure.

Proposition 2.19 *For every $Y \in \mathfrak{X}_0(M)$, the space*

$$\mathcal{M}(Y) := \{(l, \gamma) \mid l \in [0, +\infty), \ \gamma : [0, l] \to M,$$
$$\dot{\gamma}(s) + (\nabla f)(\gamma(s)) + Y_l(s, \gamma(s)) = 0\}$$

has the structure of a smooth manifold with boundary such that it is diffeomorphic to $\mathcal{M}(f, g)$. In particular:

$$\dim \mathcal{M}(Y) = n + 1 \,.$$

Proof Let $Y \in \mathfrak{X}_0(M)$. The space $\mathcal{M}(Y)$ can be reformulated as

$$\mathcal{M}(Y) = \{(l, \gamma) \mid (Y, l, \gamma) \in \widetilde{\mathcal{M}}\} \,.$$

This implies that $\psi\left(\{Y\} \times \mathcal{M}(f, g)\right) = \{Y\} \times \mathcal{M}(Y)$.

Since $\{Y\} \times \mathcal{M}(f, g)$ is a smooth submanifold with boundary of $\mathfrak{X}_0(M) \times \mathcal{M}(f, g)$, the space $\{Y\} \times \mathcal{M}(Y)$ is a smooth submanifold of $\widetilde{\mathcal{M}}$ diffeomorphic to $\{Y\} \times \mathcal{M}(f, g)$ by definition of the smooth structure on $\widetilde{\mathcal{M}}$. Forgetting about the factor $\{Y\}$ shows the claim. $\qquad\qquad\qquad\qquad\qquad\qquad\qquad\qquad\square$

For $l \geq 0$, let $\widetilde{\mathcal{M}}_l := \left\{(Y, \lambda, \gamma) \in \widetilde{\mathcal{M}} \,\middle|\, \lambda = l\right\}$.

Theorem 2.20   1. *The map $E : \widetilde{\mathcal{M}} \to M^2$, $(Y, (l, \gamma)) \mapsto (\gamma(0), \gamma(l))$, is of class C^{n+1}.*
   2. *Its restriction to $\widetilde{\mathcal{M}}_l$ is a submersion for every $l > 0$.*

Proof   1. By definition of the smooth structure on $\widetilde{\mathcal{M}}$, we need to show that the map

$$E \circ \psi : \mathfrak{X}_0(M) \times \mathcal{M}(f, g) \to M^2$$

is of class C^{n+1} and that its restriction to the interior of its domain is a submersion. By definition of the smooth structure on $\mathcal{M}(f, g)$, this is in turn equivalent to showing that the map

$$E_0 : \mathfrak{X}_0(M) \times [0, +\infty) \times M \to M^2 \,, \quad E_0 := E \circ \psi \circ \left(\mathrm{id}_{\mathfrak{X}_0(M)} \times \varphi^{-1}\right) \,,$$

is of class C^{n+1} and that its restriction to $\mathfrak{X}_0(M) \times [0, +\infty) \times M$ is a submersion of class C^{n+1}.

For this purpose, we want to write down the map E_0 more explicitly. Let ϕ^f denote the negative gradient flow of (f, g). For $(Y, l, x) \in \mathfrak{X}_0(M) \times [0, +\infty) \times M$, we compute

$$\left(\psi \circ \varphi^{-1}\right)(Y, l, x) = \psi\left(Y, \left(l, \left(s \mapsto \phi_s^f(x), \ s \in [0, l]\right)\right)\right)$$
$$= \left(Y, \left(l, \left(s \mapsto \phi_{s, \frac{1}{2}}^{Y, l}\left(\phi_{\frac{1}{2}}^f(x)\right), \ s \in [0, l]\right)\right)\right).$$

Using this result, we further derive

$$E_0(Y, l, x) = E\left(Y, \left(l, \left(s \mapsto \phi_{s, \frac{1}{2}}^{Y, l}\left(\phi_{\frac{1}{2}}^f(x)\right), \ s \in [0, l]\right)\right)\right)$$
$$= \left(\left(\phi_{0, \frac{1}{2}}^{Y, l} \circ \phi_{\frac{1}{2}}^f\right)(x), \left(\phi_{l, \frac{1}{2}}^{Y, l} \circ \phi_{\frac{1}{2}}^f\right)(x)\right). \qquad\qquad (2.16)$$

Consider the map

$$\sigma : \mathfrak{X}_0(M) \times [0, +\infty) \times \mathbb{R} \times M \to T\mathfrak{X}_0(M) \times T[0, +\infty) \times T\mathbb{R} \times TM \,,$$
$$(Y, l, t, x) \mapsto (0, 0, 0, Y(l, t, x)) \,.$$

The evaluation map defined on a mapping space consisting of maps of class C^{n+1} is itself of class C^{n+1}, see [Irw80, Corollary B.16]. Therefore, σ is a vector field of class C^{n+1} on $\mathfrak{X}_0(M) \times [0, +\infty) \times \mathbb{R} \times M$ and has a flow of class C^{n+1} (defined for non-negative time values):

$$\phi^\sigma : [0, +\infty) \times \mathfrak{X}_0(M) \times [0, +\infty) \times \mathbb{R} \times M \to \mathfrak{X}_0(M) \times [0, +\infty) \times \mathbb{R} \times M ,$$
$$(s, Y, l, t, x) \mapsto \phi_s^\sigma(Y, l, t, x) .$$

Let $\pi : \mathfrak{X}_0(M) \times [0, +\infty) \times \mathbb{R} \times M \to M$ denote the projection onto the last factor and put

$$\tilde{\phi}^\sigma : [0, +\infty) \times \mathfrak{X}_0(M) \times [0, +\infty) \times \mathbb{R} \times M \to M , \quad \tilde{\phi}^\sigma := \pi \circ \phi^\sigma .$$

By definition of σ, the uniqueness of integral curves of vector fields implies that

$$\tilde{\phi}^\sigma(s, Y, l, t, x) = \phi_s^{Y,l}(t, x) .$$

Therefore, the map $(s, Y, l, t, x) \mapsto \phi_s^{Y,l}(t, x)$ is of class C^{n+1}. Applying this result and the fact that ϕ^f is a smooth map to (2.16), one sees that E_0 is a composition of maps of class C^{n+1} and therefore itself of class C^{n+1}.

2. The claim is equivalent to $E_0|_{\mathfrak{X}_0(M) \times \{l\} \times M}$ being a submersion for every $l > 0$. We can w.l.o.g. assume that $l = 2$ and put $\phi_{s,t}^Y := \phi_{s,t}^{Y,2}$ for all s and t. This simplifies the notation since by (2.16):

$$E_{0,2}(Y, x) := E_0(Y, 2, x) = \left(\left(\phi_{0,1}^Y \circ \phi_1^f \right)(x), \left(\phi_{2,1}^Y \circ \phi_1^f \right)(x) \right) .$$

Moreover, for every $v \in T_x M$ the following holds:

$$\begin{aligned} &\left(DE_{0,2} \right)_{(Y,x)} [(0, v)] \\ &= \left(\left(D\phi_{0,1}^Y \right)_{\phi_1^f(x)} \left[\left(D\phi_1^f \right)_x [v] \right], \left(D\phi_{2,1}^Y \right)_{\phi_1^f(x)} \left[\left(D\phi_1^f \right)_x [v] \right] \right) . \end{aligned} \tag{2.17}$$

Let $(v_1, v_2) \in T_{E_{0,2}(Y,x)} M^2$. We need to find an element of $T_{(Y,x)}(\mathfrak{X}_0(M) \times M)$ which maps to (v_1, v_2) under $(DE_{0,2})_{(Y,x)}$.
Since flow maps are diffeomorphisms, there is a unique $w_0 \in T_{\phi_1^f(x)} M$ with

$$\left(D\phi_{0,1}^Y \right)_{\phi_1^f(x)} [w_0] = v_1 ,$$

and a unique $v_0 \in T_x M$ with $\left(D\phi_1^f \right)_x [v_0] = w_0$. For this choice of tangent vectors, Eq. (2.17) yields:

$$\left(DE_{0,2} \right)_{(Y,x)} [(0, v_0)] = \left(v_1, \left(D\phi_{2,1}^Y \right)_{\phi_1^f(x)} [w_0] \right) .$$

To show the claim, we therefore need to find $Z \in T_Y \mathfrak{X}_0(M)$ with

$$\left(DE_{0,2}\right)_{(Y,x)}[(Z,0)] = \left(0, v_2 - \left(D\phi_{2,1}^Y\right)_{\phi_1^f(x)}[w_0]\right),$$

since this would imply $\left(DE_{0,2}\right)_{(Y,x)}(Z, v_0) = (v_1, v_2)$. We will construct such a Z using isotopy theory.

Put $y_0 := \left(\phi_{2,1}^Y \circ \phi_1^f\right)(x)$ and pick a smooth curve $\alpha : [0, 1] \to M$ with

$$\alpha(0) = y_0, \qquad \dot{\alpha}(0) = v_2 - \left(D\phi_{2,1}^Y\right)_{\phi_1^f(x)}[w_0] \in T_{y_0}M. \qquad (2.18)$$

Let $y_1 := \alpha(1)$. We can view α as a smooth isotopy from y_0 to y_1, where we see y_0 and y_1 as zero-dimensional submanifolds of M.

By the isotopy extension theorem, see [Hir76, Sect. 8.1], we can extend α to a smooth diffeotopy

$$F : [0, 1] \times M \to M.$$

Moreover, we can choose F to have its support in a small neighborhood of $\alpha([0, 1])$ and especially such that

$$F\left(t, \phi_{s,1}^Y\left(\phi_1^f(x)\right)\right) = \phi_{s,1}^Y\left(\phi_1^f(x)\right) \quad \forall s \in [0, \tfrac{4}{3}]. \qquad (2.19)$$

For every $s \in [0, 2]$ we then define a time-dependent tangent vector at $\phi_{s,1}^{Y_2}\left(\phi_1^f(x)\right)$ by

$$Z_2\left(s, \phi_{s,1}^{Y_2}\left(\phi_1^f(x)\right)\right) := \frac{\partial}{\partial t} F\left(t, \phi_{s,1}^{Y_2}\left(\phi_1^f(x)\right)\right)\Big|_{t=0},$$

where $\frac{\partial}{\partial t}$ denotes the derivative of F in the $[0, 1]$-direction. Using (2.18) and (2.19), this yields:

$$\begin{aligned} Z_2\left(s, \phi_{s,1}\left(\phi_1^f(x)\right)\right) &= 0 \ \forall s \in [0, \tfrac{4}{3}], \\ Z_2\left(2, \phi_{2,1}\left(\phi_1^f(x)\right)\right) &= v_2 - \left(D\phi_{2,1}^Y\right)_{\phi_1^f(x)}[w_0]. \end{aligned} \qquad (2.20)$$

We can extend Z_2 to a smooth, time-dependent vector field fulfilling

$$Z_2(t, x) = 0 \ \forall t \in [0, \tfrac{4}{3}], \quad x \in M. \qquad (2.21)$$

Such a vector field satisfying (2.21) can in turn be extended to a parametrized vector field $Z \in \mathfrak{X}_0(M)$ satisfying

$$Z(2, t, x) = Z_2(t, x) \qquad \forall t \in \mathbb{R}, \ x \in M.$$

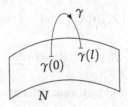

Fig. 2.4 An element of $\mathcal{M}(Y, N)$

(Note that by definition of $\mathfrak{X}_0(M)$, condition (2.21) is required for this extendability of Z_2.) For such a particular choice of Z we obtain

$$\left(DE_{2,0}\right)_{(Y,x)}(Z, 0) = \left(\left(D\tilde{\phi}_{0,1}^{\sigma}\right)_{(Y,\phi_f^1(x))}[(Z, 0)], \left(D\tilde{\phi}_{2,1}^{\sigma}\right)_{(Y,\phi_f^1(x))}[(Z, 0)]\right)$$

$$= \left(Z_2\left(0, \phi_{0,1}^Y\left(\phi_1^f(x)\right)\right), Z_2\left(2, \phi_{2,1}^Y\left(\phi_1^f(x)\right)\right)\right)$$

$$\overset{(2.20)}{=} \left(0, v_2 - \left(D\phi_{2,1}^Y\right)_{\phi_1^f(x)}[w_0]\right).$$

Hence we have shown that for any choice of v_1 and v_2 there is

$$(Z, v_0) \in T_{(Y,x)}\left(\mathfrak{X}_0(M) \times M\right)$$

with $\left(DE_{0,2}\right)_{(Y,x)}[(Z, v_0)] = (v_1, v_2)$, which shows the claim. $\square$

Remark 2.21 Note that for every $(Y, (0, \gamma)) \in \widetilde{\mathcal{M}}$, we obtain

$$E(Y, 0, \gamma) = (\gamma(0), \gamma(0)).$$

So the restriction of E to $\widetilde{\mathcal{M}}_0$ is *not* submersive if $n > 0$.

In the same way as Theorem 2.14 is derived from Theorem 2.13, we can deduce the following statement from Theorem 2.20 and the Sard–Smale theorem. The attentive reader will have no problem providing a detailed proof.

Theorem 2.22 *Let $N \subset M^2$ be a submanifold. There is a generic set $\mathcal{G} \subset \mathfrak{X}_0(M)$, such that for every $Y \in \mathcal{G}$ the space*

$$\mathcal{M}(Y, N) := \{(l, \gamma) \in \mathcal{M}(Y) \mid (\gamma(0), \gamma(l)) \in N\}$$

is a submanifold with boundary of $\mathcal{M}(Y)$ of class C^{n+1} with

$$\dim \mathcal{M}(Y, N) = n + 1 - \operatorname{codim}_{M^2} N = \dim N + 1 - n.$$

See Fig. 2.4 for an illustration of a space of the form $\mathcal{M}(Y, N)$.

Chapter 3
Nonlocal Generalizations

In this chapter we derive a more general, and in a certain sense nonlocal, version of the transversality results from the previous chapter. More precisely, we want to derive a transversality theorem, our Theorem 3.10, in which:

- all three types of trajectories from the previous chapter are considered at once,
- the perturbations may depend on the interval length parameters of *all* of the finite-length trajectories involved,
- the limiting behaviour of the perturbations for length parameters becoming infinitely large can be controlled a priori.

The first bullet point makes it possible to consider more sophisticated constructions of moduli spaces of trajectories with transverse intersections. While we have studied individual trajectories whose endpoint evaluation maps intersect submanifolds of M in the semi-infinite case or of M^2 in the finite-length case, we want to consider families of trajectories whose associated product of endpoint evaluation maps intersects a submanifold of M^N for a suitable $N \in \mathbb{N}$. This construction will be used in the discussion of perturbed Morse ribbon trees in Chap. 4.

The second and third bullet point give us the possibility to control the compactifications of these moduli spaces in a convenient way, as we will see for the special case of Morse ribbon trees in Chaps. 5 and 6.

3.1 Parametrized Perturbation Spaces

We start by defining generalizations of the perturbation spaces $\mathfrak{X}_\pm(M)$ and $\mathfrak{X}_0(M)$ from the previous section. These will be needed to construct perturbations fulfilling the condition in the second bullet.

© Springer International Publishing AG, part of Springer Nature 2018
S. Mescher, *Perturbed Gradient Flow Trees and A∞-algebra Structures in Morse Cohomology*, Atlantis Studies in Dynamical Systems 6,
https://doi.org/10.1007/978-3-319-76584-6_3

For $k > 0, j \in \{1, 2, \ldots, k\}, \lambda \geq 0, Y \in C^{n+1}\left([0, +\infty)^k \times (-\infty, 0] \times M, TM\right)$
we define:

$$c_j(\lambda, Y) \in C^{n+1}\left([0, +\infty)^{k-1} \times (-\infty, 0] \times M, TM\right) ,$$

$$c_j(\lambda, Y)(l_1, \ldots, l_{k-1}, t, x) := Y\left(l_1, \ldots, l_{j-1}, \lambda, l_j, \ldots, l_{k-1}, t, x\right) .$$

(3.1)

In other words, $c_j(\lambda, Y)$ is the contraction of Y obtained by inserting λ into the jth component.

Definition 3.1 For $k \in \mathbb{N}_0$ we recursively define a space $\mathfrak{X}_-(M, k)$ by putting

$$\mathfrak{X}_-(M, 0) := \mathfrak{X}_-(M) ,$$
$$\mathfrak{X}_-(M, k)$$
$$:= \left\{ Y \in C^{n+1}\left([0, +\infty)^k \times (-\infty, 0] \times M, TM\right) \, \middle| \, Y\left(\vec{l}, \cdot, \cdot\right) \in \mathfrak{X}_-(M) \; \forall \vec{l} \in [0, +\infty)^k, \right.$$
$$\left. \lim_{\lambda \to +\infty} c_j(\lambda, Y) \text{ exists in } \mathfrak{X}_-(M, k-1) \; \forall j \in \{1, 2, \ldots, k\} \right\} .$$

In strict analogy with the spaces $\mathfrak{X}_-(M, k)$ we define

$$\mathfrak{X}_+(M, k) \subset C^{n+1}\left([0, +\infty)^k \times [0, +\infty) \times M, TM\right)$$

for every $k \in \mathbb{N}_0$ with $\mathfrak{X}_+(M, 0) = \mathfrak{X}_+(M)$.

We next want to define analogous perturbation spaces $\mathfrak{X}_0(M, k)$ for finite-length trajectories for every $k \in \mathbb{N}_0$. This requires some additional preparations, namely we need to introduce straightforward generalizations of the map split_0 from (2.9).
For $k \in \mathbb{N}_0, l \geq 3$ and a map

$$Z : [0, +\infty)^k \times [0, +\infty) \times [0, +\infty) \times M \to TM , \quad \left(\vec{l}, l, t, x\right) \mapsto Z\left(\vec{l}, l, t, x\right) ,$$

we define maps

$$s_k(l, Z) : [0, +\infty)^k \times [0, +\infty) \times M \to TM, \quad \left(\vec{l}, t, x\right) \mapsto \begin{cases} Z\left(\vec{l}, l, t, x\right) & \text{if } t \in [0, 1] , \\ 0 \in T_x M & \text{if } t > 1 , \end{cases}$$

$$e_k(l, Z) : [0, +\infty)^k \times (-\infty, 0] \times M \to TM, \quad \left(\vec{l}, t, x\right) \mapsto \begin{cases} Z\left(\vec{l}, l, t+l, x\right) & \text{if } t \in [-1, 0] , \\ 0 \in T_x M & \text{if } t < -1 . \end{cases}$$

We then put:

$$\text{split}_k(l, Z) := (s_k(l, Z), e_k(l, Z)) .$$

(3.2)

Definition 3.2 For every $k \in \mathbb{N}_0$ we define a space $\mathfrak{X}_0(M, k)$ recursively by putting

$$\mathfrak{X}_0(M, 0) := \mathfrak{X}_0(M),$$
$$\mathfrak{X}_0(M, k) := \Big\{ Y \in C^{n+1} \left([0, +\infty)^k \times [0, +\infty) \times [0, +\infty) \times M, TM \right)$$
$$\Big| \; Y\left(\vec{l}, \cdot, \cdot, \cdot \right) \in \mathfrak{X}_0(M) \; \forall \vec{l} \in [0, +\infty)^k, \tag{3.3}$$
$$\lim_{\lambda \to +\infty} c_j(\lambda, Y) \text{ exists in } \mathfrak{X}_0(M, k-1) \; \forall j \in \{1, 2, \ldots, k\},$$
$$\lim_{l \to +\infty} \mathrm{split}_k(l, Y) \text{ exists in } \mathfrak{X}_+(M, k) \times \mathfrak{X}_-(M, k) \Big\},$$

where the maps $c_1, \ldots, c_k$ are defined in strict analogy with (3.1).

Remark 3.3 Arguing as in Remark 2.18, one observes that for any

$$Y \in C^{n+1} \left([0, +\infty)^k \times [0, +\infty) \times [0, +\infty) \times M, TM \right)$$

which fulfills condition (3.3) and for which the limit $Y_0 := \lim_{l \to +\infty} \mathrm{split}_k(l, Y)$ exists pointwise, Y_0 is a pair of parametrized vector fields of class C^{n+1} and already lies in $\mathfrak{X}_+(M, k) \times \mathfrak{X}_-(M, k)$.

One checks that all the spaces $\mathfrak{X}_\pm(M, k)$ and $\mathfrak{X}_0(M, k)$ are closed linear subspaces of Banach spaces, hence themselves Banach spaces.

3.2 Background Perturbations and Convergence

Before we provide the nonlocal transversality theorem we are longing for, we still need to define another analytic notion, namely the notion of background perturbations. These perturbations are introduced with regard to the limiting behaviour of perturbations. Technically, this method will result in the consideration of affine subspaces of the spaces $\mathfrak{X}_\pm(M, k)$ and $\mathfrak{X}_0(M, k)$.

Definition 3.4 For $k \in \mathbb{N}$ the spaces $\mathfrak{X}_\pm^{\mathrm{back}}(M, k)$ and $\mathfrak{X}_+^{\mathrm{back}}(M, k)$ are defined by

$$\mathfrak{X}_\pm^{\mathrm{back}}(M, k) := \Big\{ \mathbf{X} = (X_1, \ldots, X_k) \in \mathfrak{X}_\pm(M, k-1)^k$$
$$\Big| \lim_{\lambda \to \infty} c_i \left(\lambda, X_j \right) = \lim_{\lambda \to \infty} c_{j-1} \left(\lambda, X_i \right) \; \forall i < j \in \{1, 2, \ldots, k\} \Big\}. \tag{3.4}$$

$\mathfrak{X}_-^{\mathrm{back}}(M, k)$ (resp. $\mathfrak{X}_+^{\mathrm{back}}(M, k)$) is called the space of *negative (resp. positive) k-parametrized background perturbations on* M. For $\mathbf{X} = (X_1, \ldots, X_k) \in \mathfrak{X}_\pm^{\mathrm{back}}(M, k)$ define

$$\mathfrak{X}_\pm(M, k, \mathbf{X}) := \Big\{ Y \in \mathfrak{X}_\pm(M, k) \Big| \lim_{\lambda \to +\infty} c_j(\lambda, Y) = X_j \in \mathfrak{X}_\pm(M, k-1) \; \forall j \in \{1, \ldots, k\} \Big\}.$$

Lemma 3.5 *For all* $k \in \mathbb{N}$ *and* $\mathbf{X} \in \mathfrak{X}_{\pm}^{back}(M, k)$, *the space* $\mathfrak{X}_{\pm}(M, k, \mathbf{X})$ *is a Banach submanifold of* $C^{n+1}\left([0, +\infty)^k \times \mathbb{R}_{\pm} \times M, TM\right)$, *where* $\mathbb{R}_{+} := [0, +\infty)$, $\mathbb{R}_{-} := (-\infty, 0]$.

Proof Since every $\mathbf{X} \in \mathfrak{X}_{\pm}^{back}(M, k)$ satisfies condition (3.4) by definition, one can easily find $X \in \mathfrak{X}_{\pm}(M, k)$ satisfying

$$\lim_{\lambda \to +\infty} c_j(\lambda, X) = X_j \, .$$

Given such an X, we can reformulate $\mathfrak{X}_{\pm}(M, k, \mathbf{X})$ as an affine subspace

$$\mathfrak{X}_{\pm}(M, k, \mathbf{X}) = X + \mathfrak{X}_{\pm}(M, k, 0) \, . \tag{3.5}$$

where $0 \in \mathfrak{X}_{\pm}^{back}(M, k)$ denotes the background perturbation consisting only of vector fields which are constant to zero. It is easy to check that the space $\mathfrak{X}_{\pm}(M, k, 0)$ is a closed linear subspace of $\mathfrak{X}_{\pm}(M, k)$.

Hence by (3.5), the space $\mathfrak{X}_{\pm}(M, k, \mathbf{X})$ is a closed affine subspace and therefore a Banach submanifold of $C^{n+1}\left([0, +\infty)^k \times \mathbb{R}_{\pm} \times M, TM\right)$. $\qquad \square$

For the definition of background perturbations for finite-length trajectories, we additionally need an analogous condition to (3.4) involving the maps split_k.

Definition 3.6 1. For every $k \in \mathbb{N}_0$ *the space of* k-*parametrized finite-length background perturbations is defined as*

$$\mathfrak{X}_0^{back}(M, k) := \Big\{(X_1, \dots, X_k, X^+, X^-) \in (\mathfrak{X}_0(M, k-1))^k \times \mathfrak{X}_+(M, k) \times \mathfrak{X}_-(M, k)$$

$$\Big| \lim_{\lambda \to +\infty} c_i\left(\lambda, X_j\right) = \lim_{\lambda \to +\infty} c_{j-1}\left(\lambda, X_i\right) \ \forall i < j \in \{1, \dots, k\}, \tag{3.6}$$

$$\lim_{\lambda \to +\infty} \text{split}_{k-1}(\lambda, X_j) = \lim_{\lambda \to +\infty} \left(c_j(\lambda, X^+), c_j(\lambda, X^-)\right)\Big\} \, . \tag{3.7}$$

2. For $\mathbf{X} = (X_1, \dots, X_k, X^+, X^-) \in \mathfrak{X}_0^{back}(M, k)$, we define

$$\mathfrak{X}_0(M, k, \mathbf{X}) := \Big\{ Y \in \mathfrak{X}_0(M, k) \ \Big| \ \lim_{\lambda \to \infty} c_j(\lambda, Y) = X_j \ \forall j \in \{1, 2, \dots, k\},$$

$$\lim_{l \to \infty} \text{split}_k(l, Y) = \left(X^+, X^-\right)\Big\} \, .$$

Lemma 3.7 *For all* $k \in \mathbb{N}_0$ *and* $\mathbf{X} \in \mathfrak{X}_0^{back}(M, k)$, *the space* $\mathfrak{X}_0(M, k, \mathbf{X})$ *is a Banach submanifold of* $\mathfrak{X}_0(M, k)$.

Proof We show the claim in analogy with the proof of Lemma 3.5. Since $\mathbf{X} \in \mathfrak{X}_0^{back}(M, k)$ is defined to satisfy conditions (3.6) and (3.7), we can find a parametrized vector field $X \in \mathfrak{X}_0(M, k)$ satisfying

$$\lim_{\lambda \to +\infty} c_j\left(\lambda, X\right) = X_j \ \forall j \in \{1, 2, \dots, k\}, \quad \text{and} \quad \lim_{\lambda \to +\infty} \text{split}_k(\lambda, X) = (X^+, X^-) \, .$$

Given such an X, we can reformulate $\mathfrak{X}_0(M, k, \mathbf{X})$ as a closed affine subspace

$$\mathfrak{X}_0(M, k, \mathbf{X}) = X + \mathfrak{X}_0(M, k, 0)$$

where $0 \in \mathfrak{X}_0^{back}(M, k)$ denotes the background perturbations consisting only of vector field which are constantly zero. The rest of the proof is carried out along the same lines as the one of Lemma 3.5. $\qquad\square$

Lemma 3.8 *For* $k \geq 1$, $\mathbf{X}_- \in \mathfrak{X}_-^{back}(M, k)$, $\mathbf{X}_+ \in \mathfrak{X}_+^{back}(M, k)$ *and* $\mathbf{X}_0 \in \mathfrak{X}_0^{back}$ (M, k), *consider the maps:*

$$p_k^- : [0, +\infty)^k \times \mathfrak{X}_-(M, k, \mathbf{X}_-) \to \mathfrak{X}_-(M),$$
$$p_k^+ : [0, +\infty)^k \times \mathfrak{X}_+(M, k, \mathbf{X}_+) \to \mathfrak{X}_+(M),$$
$$p_k^0 : [0, +\infty)^k \times \mathfrak{X}_0(M, k, \mathbf{X}_0) \to \mathfrak{X}_0(M),$$

which are all three defined by

$$\left(\vec{l}, Y\right) \mapsto Y\left(\vec{l}, \cdot\right).$$

All three maps are of class C^{n+1}. *Moreover, for every* $\vec{l} \in [0, +\infty)^k$, *the maps*

$$p_k^\pm\left(\vec{l}, \cdot\right) : \mathfrak{X}_\pm(M, k, \mathbf{X}_\pm) \to \mathfrak{X}_\pm(M),$$
$$p_k^0\left(\vec{l}, \cdot\right) : \mathfrak{X}_0(M, k, \mathbf{X}_0) \to \mathfrak{X}_0(M),$$

are surjective submersions.

Proof The $(n + 1)$-fold differentiability of the maps follows again since evaluation maps considered on spaces of maps of class C^{n+1} are themselves of class C^{n+1}. Moreover, one checks that for any $\vec{l} \in [0, +\infty)^k$ the maps $p_k^\pm\left(\vec{l}, \cdot\right)$ and $p_k^0\left(\vec{l}, \cdot\right)$ are surjective and continuous linear maps between Banach spaces, hence they are submersions. $\qquad\square$

In the notation of Lemma 3.8, we will occasionally write

$$Y_\pm\left(\vec{l}\right) := p_k^\pm\left(\vec{l}, Y_\pm\right), \quad Y_0\left(\vec{l}\right) := p_k^0\left(\vec{l}, Y_0\right),$$

for $\vec{l} \in [0, +\infty)^k$, $Y_\pm \in \mathfrak{X}_\pm(M, k)$ and $Y_0 \in \mathfrak{X}_0(M, k)$.

The ultimate aim of this section is to derive a nonlocal transversality theorem which considers perturbed negative semi-infinite, positive semi-infinite and finite-length Morse trajectories all at once. To improve the clarity of the exposition, we introduce some additional notation:

Definition 3.9 For $k_1, k_2, k_3 \in \mathbb{N}_0$ with $k_2 > 0$ define

$$\mathfrak{X}(k_1, k_2, k_3) := (\mathfrak{X}_-(M, k_2))^{k_1} \times (\mathfrak{X}_0(M, k_2 - 1))^{k_2} \times (\mathfrak{X}_+(M, k_2))^{k_3} .$$

We call $\mathfrak{X}(k_1, k_2, k_3)$ *the space of perturbations of type* (k_1, k_2, k_3). We further define

$$\mathfrak{X}^{\text{back}}(k_1, k_2, k_3) := \left(\mathfrak{X}_-^{\text{back}}(M, k_2)\right)^{k_1} \times \left(\mathfrak{X}_0^{\text{back}}(M, k_2 - 1)\right)^{k_2} \times \left(\mathfrak{X}_+^{\text{back}}(M, k_2)\right)^{k_3}$$

and call it *the space of background perturbations of type* (k_1, k_2, k_3).
　　We write $\mathbf{X} \in \mathfrak{X}(k_1, k_2, k_3)$ as

$$\mathbf{X} := \left(\mathbf{X}_1^-, \mathbf{X}_2^-, \ldots, \mathbf{X}_{k_1}^-, \mathbf{X}_1^0, \mathbf{X}_2^0, \ldots, \mathbf{X}_{k_2}^0, \mathbf{X}_1^+, \mathbf{X}_2^+, \ldots, \mathbf{X}_{k_3}^+\right)$$

with $\mathbf{X}_i^- \in \mathfrak{X}_-^{\text{back}}(M, k_2)$ for every $i \in \{1, 2, \ldots, k_1\}$, $\mathbf{X}_j^0 \in \mathfrak{X}_0^{\text{back}}(M, k_2 - 1)$ for every $j \in \{1, 2, \ldots, k_2\}$ and $\mathbf{X}_k^+ \in \mathfrak{X}_+^{\text{back}}(M, k_2)$ for every $k \in \{1, 2, \ldots, k_3\}$, and define

$$\mathfrak{X}(k_1, k_2, k_3, \mathbf{X}) := \prod_{i=1}^{k_1} \mathfrak{X}_- \left(M, k_2, \mathbf{X}_i^-\right) \times \prod_{j=1}^{k_2} \mathfrak{X}_0 \left(M, k_2 - 1, \mathbf{X}_j^0\right) \times \prod_{k=1}^{k_3} \mathfrak{X}_+ \left(M, k_2, \mathbf{X}_k^+\right) .$$

3.3　A Nonlocal Transversality Theorem

Having collected all necessary ingredients in the first two sections we are finally able to state and prove the aforementioned nonlocal transversality theorem.

Theorem 3.10 *Let* $k_1, k_2, k_3 \in \mathbb{N}_0$. *Let* V *be a smooth submanifold of* $(0, +\infty)^{k_2}$ *and* N *be a smooth submanifold of* $M^{k_1 + 2k_2 + k_3}$. *If* $k_2 > 0$, *then let additionally* $\mathbf{X} \in \mathfrak{X}^{back}(k_1, k_2, k_3)$. *There is a generic subset* $\mathcal{G} \subset \mathfrak{X}(k_1, k_2, k_3, \mathbf{X})$ *if* $k_2 > 0$ *and* $\mathcal{G} \subset \mathfrak{X}(k_1, 0, k_3)$ *if* $k_2 = 0$, *such that for every* $\mathbf{Y} = \left(Y_1^-, \ldots, Y_{k_1}^-, Y_1, \ldots, Y_{k_2}, Y_1^+, \ldots, Y_{k_3}^+\right) \in \mathcal{G}$ *the space*

$$\mathcal{M}_\mathbf{Y}(x_1, x_2, \ldots, x_{k_1}, y_1, y_2, \ldots, y_{k_3}, (k_1, k_2, k_3), V, N)$$
$$:= \Big\{ \left(\gamma_1^-, \ldots, \gamma_{k_1}^-, (l_1, \gamma_1), \ldots, (l_{k_2}, \gamma_{k_2}), \gamma_1^+, \ldots, \gamma_{k_3}^+\right) \mid (l_1, l_2, \ldots, l_{k_2}) \in V ,$$
$$\gamma_i^- \in W^- \left(x_i, Y_i^-(l_1, l_2, \ldots, l_{k_2})\right) \ \forall i \in \{1, 2, \ldots, k_1\} ,$$
$$\gamma_j^+ \in W^- \left(y_j, Y_j^+(l_1, l_2, \ldots, l_{k_2})\right) \ \forall j \in \{1, 2, \ldots, k_3\} ,$$
$$(l_j, \gamma_j) \in \mathcal{M} \left(Y_j(l_1, \ldots, l_{j-1}, l_{j+1}, \ldots, l_{k_2})\right) \ \forall j \in \{1, 2, \ldots, k_2\} ,$$
$$\left((E_{Y_i^-(l_1, \ldots, l_{k_2})}^-(\gamma_i^-))_{i=1,2,\ldots,k_1}, (E_{Y_j(l_1,\ldots,l_{j-1},l_{j+1},\ldots,l_{k_2})}(l_j, \gamma_j))_{j=1,2,\ldots,k_2},\right.$$
$$\left. (E_{Y_k^+(l_1,\ldots,l_{k_2})}^+(\gamma_k^+))_{k=1,2,\ldots,k_3}\right) \in N \Big\}$$

is a manifold of class C^{n+1} for all $x_1, x_2, \ldots, x_{k_1}, y_1, y_2 \ldots, y_{k_3} \in \text{Crit } f$. For every $\mathbf{Y} \in \mathcal{G}$ its dimension is given by

$$\dim \mathcal{M}_\mathbf{Y}(x_1, x_2, \ldots, x_{k_1}, y_1, y_2, \ldots, y_{k_3}, (k_1, k_2, k_3), V, N)$$

$$= \sum_{i=1}^{k_1} \mu(x_i) - \sum_{k=1}^{k_3} \mu(y_k) + (k_2 + k_3)n + \dim V - \text{codim } N .$$

Remark 3.11 This transversality theorem is nonlocal with respect to two different viewpoints. Firstly, observe that the perturbations of negative and positive semi-infinite and finite-length curves depend *on all the lengths $l_1, \ldots, l_{k_2}$ of the perturbed finite-length trajectories involved*. So the perturbations of the semi-infinite trajectories change with the interval lengths of the perturbed finite-length trajectories.

Secondly, if $V = (0, +\infty)^{k_2}$ and if N is of the form

$$N = \prod_{i=1}^{k_1} N_i^- \times \prod_{j=1}^{k_2} N_j^0 \times \prod_{k=1}^{k_3} N_k^+ ,$$

where $N_1^-, \ldots, N_{k_1}^-, N_1^+, \ldots, N_{k_3}^+ \subset M$ and $N_1^0, \ldots, N_{k_2}^0 \subset M^2$ are smooth submanifolds, then the statement of Theorem 3.10 can be derived in the spirit of the analogous results of the previous section (up to the introduction of parameters).

Since V can be an arbitrary submanifold of $(0, +\infty)^{k_2}$ and N can be an arbitrary smooth submanifold of $M^{k_1 + 2k_2 + k_3}$ and *does not need to be a product of submanifolds of the factors*, Theorem 3.10 can be regarded as a nonlocal generalization of the corresponding results in the previous section.

Proof of Theorem 3.10 For $k_2 = 0$, Theorem 2.13, part 2 of Theorem 2.15 and the Sard–Smale transversality theorem. In the following we therefore assume that $k_2 > 0$. Consider the map

$$p : V \times \mathfrak{X}(k_1, k_2, k_3, \mathbf{X}) \to \mathfrak{X}_-(M)^{k_1} \times \mathfrak{X}_0(M)^{k_2} \times \mathfrak{X}_+(M)^{k_3}$$

$$\left(\vec{l}, (Y_i^-)_{i=1,\ldots,k_1}, (Y_j^0)_{j=1,\ldots,k_2}, (Y_k^+)_{k=1,\ldots,k_3} \right) \mapsto$$

$$\left(\left(p_{k_2}^- \left(\vec{l}, Y_i^- \right) \right)_i, \left(p_{k_2-1}^0 \left((l_1, \ldots, l_{j-1}, l_{j+1}, \ldots, l_{k_2}), Y_j^0 \right) \right)_j, \left(p_{k_2}^+ \left(\vec{l}, Y_k^+ \right) \right)_k \right) .$$

Since the maps $p_{k_2}^- \left(\vec{l}, \cdot \right), p_{k_2-1}^0 \left((l_1, \ldots, l_{j-1}, l_{j+1}, \ldots, l_{k_2}), \cdot \right)$ and $p_{k_2}^+ \left(\vec{l}, \cdot \right)$ are surjective submersions of class C^{n+1} for every $\vec{l}$ by Lemma 3.8, the map p is a surjective submersion of class C^{n+1}. Consider the space

$$\widetilde{\mathcal{M}}(V) := \left\{ ((Y_1, (l_1, \gamma_1)), \ldots, (Y_{k_2}, (l_{k_2}, \gamma_{k_2}))) \in \left(\widetilde{\mathcal{M}} \right)^{k_2} \mid (l_1, \ldots, l_{k_2}) \in V \right\} .$$

One checks that $\widetilde{\mathcal{M}}(V)$ is a Banach submanifold of $\widetilde{\mathcal{M}}^{k_2}$. Moreover, it follows from part 2 of Theorem 2.20 that the restriction of $(E^0)^{k_2} : \widetilde{\mathcal{M}}^{k_2} \to M^{2k_2}$ to $\widetilde{\mathcal{M}}(V)$ is a surjective submersion of class C^{n+1}.

Let $x_1, x_2, \ldots, x_{k_1}, y_1, y_2, \ldots, y_{k_3} \in \mathrm{Crit}\, f$. Combining the above considerations of $\widetilde{\mathcal{M}}(V)$ with Theorem 2.13 and part 2 of Theorem 2.15, we obtain that the map

$$(E^-)^{k_1} \times (E^0)^{k_2} \times (E^+)^{k_3} : \prod_{i=1}^{k_1} \widetilde{W}^-(x_i, \mathfrak{X}_-) \times \widetilde{\mathcal{M}}(V) \times \prod_{k=1}^{k_3} \widetilde{W}^+(y_k, \mathfrak{X}_+) \to M^{k_1+2k_2+k_3}$$

is a surjective submersion of class C^{n+1}, where $\widetilde{\mathcal{M}}_{>0} := \left\{ (l, \gamma) \in \widetilde{\mathcal{M}} \,\middle|\, l > 0 \right\}$. We further consider the map

$$\bar{p} : \mathfrak{X}(k_1, k_2, k_3, \mathbf{X}) \times \prod_{i=1}^{k_1} \mathcal{P}_-(x_i) \times \widetilde{\mathcal{M}}(V) \times \prod_{k=1}^{k_3} \mathcal{P}_+(y_k) \to$$

$$\prod_{i=1}^{k_1} (\mathfrak{X}_-(M) \times \mathcal{P}_-(x_i)) \times \prod_{j=1}^{k_2} (\mathfrak{X}_0(M) \times \widetilde{\mathcal{M}}_{>0}) \times \prod_{k=1}^{k_3} (\mathfrak{X}_+(M) \times \mathcal{P}_+(y_k)) \,,$$

$$\left((Y_i^-)_{i=1,\ldots,k_1}, (Y_j^0)_{j=1,\ldots,k_2}, (Y_k^+)_{k=1,\ldots,k_3}, (\gamma_i^-)_{i=1,\ldots,k_1}, (Y_j, l_j, \gamma_j)_{j=1,\ldots,k_2}, (\gamma_k^+)_{k=1,\ldots,k_3} \right) \mapsto$$

$$\left(\left(p_{k_2}^- \left(\bar{l}, Y_i^- \right), \gamma_i^- \right)_{i=1,\ldots,k_1}, \left(p_{k_2}^- \left((l_1, \ldots, l_{j-1}, l_{j+1}, \ldots, l_{k_2}), Y_j^0 \right), (Y_j, l_j, \gamma_j) \right)_{j=1,\ldots,k_2}, \right.$$

$$\left. \left(p_{k_2}^+ \left(\bar{l}, Y_k^+ \right), \gamma_k^+ \right)_{k=1,\ldots,k_3} \right).$$

$\bar{p}$ can be written as a composition of the surjective submersion p and several permutations of the different components involved, so $\bar{p}$ is itself a surjective submersion. Let

$$\overline{\mathcal{W}} := \bar{p}^{-1} \left(\prod_{i=1}^{k_1} \widetilde{W}^-(x_i, \mathfrak{X}_-) \times \prod_{j=1}^{k_2} \widetilde{\mathcal{M}}_{>0} \times \prod_{k=1}^{k_3} \widetilde{W}^+(y_k, \mathfrak{X}_+) \right),$$

where for convenience we identify $\widetilde{\mathcal{M}}_{>0} \cong \{ (Y, (Y, l, \gamma)) \,|\, (Y, l, \gamma) \in \widetilde{\mathcal{M}}_{>0} \}$. Then

$$\underline{E} : \overline{\mathcal{W}} \to M^{k_1+2k_2+k_3} \,, \quad \underline{E} := \left((E^-)^{k_1} \times (E^0)^{k_2} \times (E^+)^{k_3} \right) \circ \bar{p}|_{\overline{\mathcal{W}}},$$

is well-defined, where E^-, E^0 and E^+ are defined as in the previous section. Moreover, $\underline{E}$ is a composition of surjective submersions of class C^{n+1} and therefore has the very same property. For any $\mathbf{Y} \in \mathfrak{X}(k_1, k_2, k_3, \mathbf{X})$ put

$$\overline{\mathcal{W}}_{\mathbf{Y}} := \{ ((\gamma_1^-, \ldots, \gamma_{k_1}^-), ((l_1, \gamma_1), \ldots, (l_{k_2}, \gamma_{k_2})), (\gamma_1^+, \ldots, \gamma_{k_3}^+))$$

$$\big| \, (\mathbf{Y}, ((\gamma_1^-, \ldots, \gamma_{k_1}^-), ((l_1, \gamma_1), \ldots, (l_{k_2}, \gamma_{k_2})), (\gamma_1^+, \ldots, \gamma_{k_3}^+))) \in \overline{\mathcal{W}} \}.$$

Comparing the dimensions of the respective spaces, one checks that for every choice of $\mathbf{Y} \in \mathfrak{X}(k_1, k_2, k_3, \mathbf{X})$ it holds that

$$\mathcal{M}_{\mathbf{Y}}(x_1, \ldots, x_{k_1}, y_1, \ldots, y_{k_3}, (k_1, k_2, k_3), N) = \underline{E}_{\mathbf{Y}}^{-1}(N), \qquad (3.8)$$

where $\underline{E}_{\mathbf{Y}} := \underline{E}|_{\overline{\mathcal{W}}_{\mathbf{Y}}}$. We then proceed as in the proofs of Theorems 2.14 and 2.22 to write $\underline{E}$ as a map defined on a product manifold in which $\mathfrak{X}(k_1, k_2, k_3, \mathbf{X})$ is among the factors. Applying the Sard–Smale transversality theorem yields that for generic choice of $\mathbf{Y} \in \mathfrak{X}(k_1, k_2, k_3, \mathbf{X})$, the map $\underline{E}_{\mathbf{Y}}$ is transverse to N which by (3.8) shows the claim for this particular choice of $x_1, \ldots, x_{k_1}$ and $y_1, \ldots, y_{k_3}$. The simultaneous transversality for all choices of critical points for a generic $\mathbf{Y}$ can be shown in precisely the same way as in the proof of Theorem 2.14. We omit the details.

It remains to compute the dimension of $\mathcal{M}_{\mathbf{Y}}(x_1, \ldots, x_{k_1}, y_1, \ldots, y_{k_3}, (k_1, k_2, k_3), N)$. One checks that for an arbitrary fixed $\vec{l} \in (0, +\infty)^{k_2}$ the following equation holds:

$$\dim \overline{\mathcal{W}}_{\mathbf{Y}} = \dim \left(\prod_{i=1}^{k_1} \widetilde{\mathcal{W}}^- \left(x_i, Y_i^- \left(\vec{l} \right) \right) \times \prod_{j=1}^{k_2} \widetilde{\mathcal{M}}_{>0} \left(Y_j^0 \left(l_1, \ldots, l_{j-1}, l_{j+1}, \ldots, l_{k_2} \right) \right) \right.$$
$$\left. \times \prod_{k=1}^{k_3} \widetilde{\mathcal{W}}^+ \left(y_k, Y_k^+ \left(\vec{l} \right) \right) \right) - \operatorname{codim} V .$$

Combining this identity with the dimension formula from the Sard–Smale theorem and the results of the previous section, we can conclude:

$$\dim \mathcal{M}_{\mathbf{Y}}(x_1, \ldots, x_{k_1}, y_1, \ldots, y_{k_3}, (k_1, k_2, k_3), N) = \dim \overline{\mathcal{W}} - \operatorname{codim} N$$

$$= \sum_{i=1}^{k_1} \dim W^- \left(x_i, Y_i^-(l_1, \ldots, l_{k_1}) \right) + \sum_{j=1}^{k_2} \dim \mathcal{M} \left(Y_j^0(l_1, \ldots, l_{j-1}, l_{j+1}, \ldots, l_{k_2}) \right)$$

$$+ \sum_{k=1}^{k_3} \dim W^+ \left(y_k, Y_k^+(l_1, \ldots, l_{k_2}) \right) - \operatorname{codim} V - \operatorname{codim} N$$

$$= \sum_{i=1}^{k_1} \mu(x_i) + k_2(n+1) + \sum_{k=1}^{k_3} (n - \mu(y_k)) - \operatorname{codim} V - \operatorname{codim} N$$

$$= \sum_{i=1}^{k_1} \mu(x_i) - \sum_{k=1}^{k_3} \mu(y_k) + (k_2 + k_3)n + \dim V - \operatorname{codim} N .$$

$\square$

Remark 3.12 Compared to Abouzaid's results, Theorem 3.10 generalizes Lemma 7.3 from [Abo11].

Chapter 4
Moduli Spaces of Perturbed Morse Ribbon Trees

In the remainder of of this book, we will discuss perturbed Morse ribbon trees, which can be interpreted as continuous maps from a tree to the manifold M which edgewise fulfill perturbed negative gradient flow equations. In this chapter, we will make this notion precise in terms of the constructions of Chaps. 2 and 3. Moreover, we will apply the nonlocal transversality result Theorem 3.10 to equip moduli spaces of perturbed Morse ribbon trees with the structures of finite-dimensional manifolds of class C^{n+1}.

A priori, there are two approaches to finding perturbed Morse ribbon trees. Firstly, we could start by considering continuous maps from trees to M and sorting out those whose restrictions to the edges satisfy perturbed negative gradient flow equations. Secondly, we may consider all suitable families of perturbed Morse trajectories and sort out those families which may be combined to form a continuous map from a tree to M, i.e. whose endpoints coincide in a suitable way which is made precise towards the end of Sect. 4.1.

We will follow the second approach. The condition that families of trajectories may be coupled to form continuous maps will be rephrased as an intersection condition of the family with a certain submanifold under an endpoint evaluation map of the form discussed the previous chapter.

4.1 Trees and Related Notions

We begin by giving a brief account of trees and their terminology, which we will use in the remainder of this chapter. We refrain from giving an elaborate discussion of trees and related notions from graph theory. Instead, we give a topological definition in terms of CW complexes. We will state and use several facts from graph theory without giving proofs, since this would lead us afar from the Morse-theoretic constructions we want to consider.

© Springer International Publishing AG, part of Springer Nature 2018
S. Mescher, *Perturbed Gradient Flow Trees and A∞-algebra Structures in Morse Cohomology*, Atlantis Studies in Dynamical Systems 6,
https://doi.org/10.1007/978-3-319-76584-6_4

Definition 4.1 A *tree complex* is a connected and simply connected one-dimensional CW complex. A 0-cell of a tree complex is called *external* if it is connected to precisely one 1-cell. A 0-cell which is not external is called *internal*.

We call a tree complex *ordered* if it is equipped with an ordering of its external 0-cells, i.e. a bijection

$$\{\text{external 0-cells}\} \to \{0, 1, \ldots, d\}$$

for suitable $d \in \mathbb{N}_0$.

We want to define ordered trees as equivalence classes of ordered tree complexes. For this purpose, we define a notion of isomorphisms of ordered CW complexes.

Definition 4.2 A map between two ordered tree complexes $T_1 \to T_2$ is called an *isomorphism of ordered tree complexes* if it is a cellular homeomorphism which preserves the ordering of the external 0-cells, i.e. it maps the ith external vertex of T_1 to the ith external vertex of T_2 for every $i \in \{0, 1, \ldots, d\}$.

One checks without further difficulties that this notion of isomorphism defines an equivalence relation on the class of ordered tree complexes.

Moreover, since isomorphisms of ordered tree complexes are by definition cellular maps which preserve the ordering of the external vertices, an elementary line of argument shows that these isomorphisms induce equivalence relations on the sets of 0-cells and 1-cells as well.

Definition 4.3 An isomorphism class T of ordered tree complexes is called *an ordered tree*. Isomorphism classes of 0-cells of T are called *vertices of T* and isomorphism classes of 1-cells of T are called *edges of T*. Denote the set of all vertices of T by $V(T)$ and the set of all edges of T by $E(T)$.

A vertex is called *external (internal)* if it is represented by an external (internal) 0-cell. Denote the set of all external (internal) vertices of T by $V_{ext}(T)$ (resp. $V_{int}(T)$).

An edge is called *external* if it is represented by a 1-cell which is connected to an external 0-cell. An edge is called *internal* if it is not external. Denote the set of all external (internal) edges by $E_{ext}(T)$ (resp. $E_{int}(T)$).

Since by definition isomorphisms of ordered tree complexes preserve the orderings of the external 0-cells, these orderings in turn induce orderings on the external vertices of ordered trees.

Definition 4.4 Let T be an ordered tree.

1. Denote the set of external vertices of T according to their ordering by

$$V_{ext}(T) = \{v_0(T), v_1(T), \ldots, v_d(T)\}$$

for suitable $d \in \mathbb{N}_0$. Furthermore, for $d \geq 2$ denote for $i \in \{0, 1, \ldots, d\}$ the unique external edge connected to $v_i(T)$ by $e_i(T)$. We will occasionally just write $v_0, \ldots, v_d$ if it is clear which tree we are referring to.

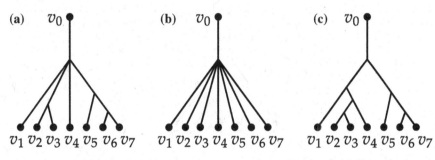

Fig. 4.1 Elements of RTree$_7$

2. The vertex $v_0(T)$ is called the *root of* T, $v_1(T)$, ..., $v_d(T)$ are called the *leaves of* T and an ordered tree is called *d-leafed* if it has precisely d leaves.
3. A vertex of T is called *n-valent*, $n \in \mathbb{N}$, (univalent, bivalent, trivalent, ...) if it is connected to precisely n different edges of T.
4. For $d \geq 2$, an ordered d-leafed tree is called *ribbon tree* if every internal vertex is at least trivalent, i.e. n-valent for some $n \geq 3$. The set of all d-leafed ribbon trees for a fixed $d \geq 2$ is denoted by RTree$_d$.
5. If e denotes the unique ordered tree having one single edge and two (external) vertices, then we will put RTree$_1 := \{e\}$.
6. A ribbon tree is called *binary tree* if every internal vertex is trivalent.

Example 4.5 Figure 4.1 depicts three different examples of 7-leafed ribbon trees. Note that tree (b) has no internal edges and a single internal vertex and that tree (c) is a binary tree.

Let $T \in$ RTree$_d$ for some $d \in \mathbb{N}$. For any $v \in V(T)$ let P_v denote the minimal subtree of T which connects the root with v, i.e. the unique subtree of T with

$$V_{ext}(P_v) = \{v_0, v\} .$$

The following statement can be shown by elementary methods of graph theory:

Lemma 4.6 *Let* $T \in$ RTree$_d$ *and* $e \in E(T)$. *Let* $v, w \in V(T)$ *be the two distinct vertices which are connected to* e. *Then either* $v \in V(P_w)$ *or* $w \in V(P_v)$.

Definition 4.7 In the situation of Lemma 4.6, assume that $v \in V(P_w)$. We then call v the *incoming vertex of* e and w the *outgoing vertex of* e. For any given edge e we further denote its incoming and outgoing vertex by

$$v_{in}(e) \quad \text{and} \quad v_{out}(e) .$$

The proof of Lemma 4.6 is elementary. Instead of providing the proof, we illustrate the situation with Fig. 4.2 which depicts an example with $d = 5$. In the picture,

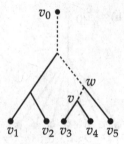

Fig. 4.2 An illustration of Lemma 4.6

the tree P_v is the subtree whose edges are the dashed edges in the picture. Here, $w \in V(P_v)$.

Remark 4.8 In this terminology, the root of a ribbon tree is always the incoming vertex of the edge it is connected to while the leaves of a ribbon tree are always the outgoing vertices of the respective edges they are connected to, i.e.

$$v_0(T) = v_{\text{in}}(e_0(T)), \quad v_i(T) = v_{\text{out}}(e_i(T)), \ \forall i \in \{1, \dots, d\}.$$

Lemma 4.9 *Let* $T \in \text{RTree}_d$ *and* $v \in V(T) \setminus \{v_0(T)\}$. *There is a unique edge* $e_v \in E(T)$ *with* $v_{\text{out}}(e_v) = v$.

Proof Since $v \neq v_0(T)$, the subtree P_v has at least one edge. By definition of P_v there is an edge $e \in E(P_v) \subset E(T)$ with $v_{\text{out}}(e) = v$.

Suppose there is an $f \in E(T)$ with $e \neq f$ and $v_{\text{out}}(f) = v$. By definition of an outgoing vertex, this would imply that there is a minimal subtree P of T with

$$V_{ext}(P) = \{v_0(T), v\}$$

and with $P \neq P_v$. Given P_v and P, one could easily construct two continuous paths $\alpha_1, \alpha_2 : [0, 1] \to T$ with $\alpha_i(0) = v_0(T)$ and $\alpha_i(1) = v$ for $i \in \{1, 2\}$ and such that

$$\text{im } \alpha_1 \subset P_v, \quad \text{im } \alpha_2 \subset P.$$

via reparametrization and concatenation, one could construct a loop in T out of α_1 and α_2 having nontrivial homotopy class. Such a loop can not exist, since T is by definition simply connected. Therefore, the claim follows.   □

After this graph-theoretic digression, we will focus our attention on maps $T \to M$. For any $T \in \text{RTree}_d$ and $e \in E(T)$ let

$$a_e : [0, 1] \to T$$

denote its attaching map. (Actually, it is an isomorphism class of attaching maps, but the meaning of this term is obvious.) We always assume that

$$a_e(0) = v_{\text{in}}(e) , \quad a_e(1) = v_{\text{out}}(e) .$$

via precomposing with the attaching maps, we can express every continuous map

$$I : T \to M$$

as a collection of maps

$$\{I \circ a_e : [0, 1] \to M \mid e \in E(T)\} .$$

Conversely, a family of maps

$$\{J_e : [0, 1] \to M \mid e \in E(T)\}$$

induces a continuous map $J : T \to M$ if and only if it satisfies the compatibility condition

$$J_e(1) = J_f(0) \quad \text{if} \quad v_{\text{out}}(e) = v_{\text{in}}(f) \quad \forall e, f \in E(T) . \tag{4.1}$$

To apply Theorem 3.10 in this context, we want to express condition (4.1) as a certain submanifold condition for a submanifold of M^N for some $N \in \mathbb{N}$, which will lead us to the definition of T-diagonals after some minor preparations.

For the rest of this book, we introduce the following notation for a given $d \in \mathbb{N}$:

$$k : \text{RTree}_d \to \mathbb{N}_0 , \quad T \mapsto |E_{int}(T)| .$$

4.2   Tree Diagonals

Let $T \in \text{RTree}_d$ and consider the product manifold $M^{1+2k(T)+d}$. For convenience, we label its factors by the edges of T, and write a point in $M^{1+2k(T)+d}$ as

$$(q_0, (q_{in}^e, q_{out}^e)_{e \in E_{int}(T)}, q_1, \dots, q_d) \in M^{1+2k(T)+d}$$

We actually need an ordering on $E_{int}(T)$ to make this well-defined, but since the ordering is irrelevant for what follows, we silently *assume that we have chosen an arbitrary ordering on $E_{int}(T)$ throughout the following discussion and leave it out of the notation.*

Let $w_0 \in V(T)$ be given by $w_0 = v_{\text{out}}(e_0(T))$. (Recall that $v_{\text{in}}(e_0(T)) = v_0(T)$.) Define

$$\begin{aligned}
\Delta_{w_0} := \{ &(q_0, (q_{in}^e, q_{out}^e)_{e \in E_{int}(T)}, q_1, \dots, q_d) \in M^{1+2k(T)+d} \\
&\mid q_0 = q_{in}^e \text{ for every } e \in E_{int}(T) \text{ with } v_{in}(e) = w_0 , \\
&\quad q_0 = q_j \text{ for every } j \in \{1, \dots, d\} \text{ with } v_{in}(e_j(T)) = w_0 \} .
\end{aligned}$$

Let $v \in V_{int}(T) \setminus \{w_0\}$. By Lemma 4.9 there is a *unique* internal edge $e_v \in E_{int}(T)$ with

$$v_{out}(e_v) = v .$$

For any such v and associated e_v we define

$$\Delta_v := \{(q_0, (q_{in}^e, q_{out}^e)_{e \in E_{int}(T)}, q_1, \ldots, q_d) \in M^{1+2k(T)+d} \tag{4.2}$$
$$| \; q_{out}^{e_v} = q_{in}^e \; \text{ for every } \; e \in E_{int}(T) \; \text{ with } \; v_{in}(e) = v ,$$
$$q_{out}^{e_v} = q_j \; \text{ for every } \; j \in \{1, \ldots, d\} \; \text{ with } \; v_{in}(e_j(T)) = v \} .$$

Definition 4.10 The T-*diagonal* $\Delta_T \subset M^{1+2k(T)+d}$ is thespace given by

$$\Delta_T := \bigcap_{v \in V_{int}(T)} \Delta_v .$$

The definition of Δ_T will be motivated in Remark 4.12.

Proposition 4.11 *For every* $T \in \mathrm{RTree}_d$ *with* $d \geq 2$, *the* T-*diagonal is a submanifold of* $M^{1+2k(T)+d}$ *with*

$$\mathrm{codim} \; \Delta_T = (k(T) + d)n . \tag{4.3}$$

Proof For every $v \in V_{int}(T)$, the space Δ_v is easily seen to be a smooth submanifold of $M^{1+2k(T)+d}$ with

$$\mathrm{codim} \; \Delta_v = |\{e \in E(T) \mid v_{in}(e) = v\}| \cdot n .$$

Furthermore, for every $V \subset V_{int}(T)$ and $v_0 \in V_{int}(T)$ with $v_0 \notin V$, the spaces Δ_{v_0} and $\bigcap_{v \in V} \Delta_v$ are smooth submanifolds and transverse to each other. Then one can build up the space Δ_T as a finite sequence of transverse intersections of submanifolds and derive that it is itself a smooth submanifold, whose codimension is given by

$$\mathrm{codim} \; \Delta_T = \sum_{v \in V_{int}(T)} \mathrm{codim} \; \Delta_v = \sum_{v \in V_{int}(T)} |\{e \in E(T) \mid v_{in}(e) = v\}| \cdot n$$
$$= |\{e \in E(T) \mid v_{in}(e) \in V_{int}(T)\}| \cdot n .$$

All that remains to do is to determine the cardinality of $\{e \in E(T) \mid v_{in}(e) \in V_{int}(T)\}$. By definition of $e_0(T)$ we know that $v_{in}(e_0(T)) = v_0 \notin V_{int}(T)$, such that $e_0(T)$ does not lie in this set. But $e_0(T)$ is the *only* edge not lying in this set. The incoming vertex of an internal edge is by definition internal. Moreover, for all external edges e with $e \neq e_0$ it holds that

$$v_{out}(e) \in \{v_1(T), \ldots, v_d(T)\} \subset V_{ext}(T) ,$$

so the incoming vertex of e has to be internal since $d \geq 2$. Consequently, we obtain

$$\text{codim } \Delta_T = |(E(T) \setminus \{e_0(T)\})| \cdot n = (k(T) + d) \cdot n .$$

□

Remark 4.12 Combining the conditions in the definitions of the different Δ_v, $v \in V_{int}(T)$, one derives that the T-diagonal is explicitly described by

$$\Delta_T := \Big\{ (q_0, (q_{in}^e, q_{out}^e)_{e \in E_{int}(T)}, q_1, \ldots, q_d) \in M^{1+2k(T)+d}$$

$$\Big| \; q_{out}^e = q_{in}^f \; \text{for all} \; e, f \in E_{int}(T) \; \text{with} \; v_{out}(e) = v_{in}(f) ,$$

$$q_{out}^e = q_j \; \text{for every} \; e \in E_{int}(T), \; j \in \{1, 2, \ldots, d\} \; \text{with} \; v_{out}(e) = v_{in}(e_j(T)) ,$$

$$q_0 = q_{in}^e \; \text{for every} \; e \in E_{int}(T) \; \text{with} \; v_{out}(e_0(T)) = v_{in}(e) \Big\} .$$

Let $\{J_e : [0, 1] \to M \mid e \in E(T)\}$ be a family of maps. If one writes down condition (4.1) in greater detail for the different types of edges, one will obtain that the family induces a continuous map $T \to M$ if and only if

$$\Big(J_{e_0}(1), (J_e(0), J_e(1))_{e \in E_{int}(T)}, J_{e_1}(0), \ldots, J_{e_d}(0) \Big) \in \Delta_T ,$$

where $e_i := e_i(T)$ for every $i \in \{0, 1, \ldots, d\}$. In other words, we have expressed condition (4.1) in terms of the T-diagonal.

4.3 Definition and Regularity of the Moduli Spaces

After all these preparations, we build the bridge to the Morse-theoretic constructions from Chaps. 2 and 3. We want to consider maps from ribbon trees to M which are not only continuous, but whose attaching maps fulfill (up to reparametrization) a perturbed negative gradient flow equation. Similar as in Remark 4.12, we will piece these maps together out of a family of maps indexed by the set of edges of the ribbon tree.

We start by describing suitable perturbation spaces. Let $d \geq 2$ and $T \in \text{RTree}_d$ be fixed.

Definition 4.13 We introduce *the space of perturbations for T* as

$$\mathfrak{X}(T) := \mathfrak{X}(1, k(T), d) ,$$

In the case that $k(T) > 0$, we consider *the space of background perturbations for T*

$$\mathfrak{X}^{\text{back}}(T) := \mathfrak{X}^{\text{back}}(1, k(T), d) ,$$

using the notation of Chap. 3. For $\mathbf{X} \in \mathfrak{X}^{\text{back}}(T)$, *the space of perturbations for T with background perturbation* $\mathbf{X}$ is in that notation defined by

$$\mathfrak{X}(T, \mathbf{X}) := \mathfrak{X}(1, k(T), d, \mathbf{X}) \,.$$

We are again silently assuming that an ordering of $E_{int}(T)$ has been chosen and keep it out of the notation. We therefore write perturbations $\mathbf{Y} \in \mathfrak{X}(T)$ as

$$\mathbf{Y} = (Y_0, (Y_e)_{e \in E_{int}(T)}, Y_1, \ldots, Y_d) \,,$$

where $Y_0 \in \mathfrak{X}_-(M, k(T))$, $Y_e \in \mathfrak{X}_0(M, k(T) - 1)$ for all $e \in E_{int}(T)$ and $Y_i \in \mathfrak{X}_+(M, k(T))$ for all $i \in \{1, 2, \ldots, d\}$. In strict analogy with the notation from Definition 3.9, we further write $\mathbf{X} \in \mathfrak{X}^{\text{back}}(T)$ as

$$\mathbf{X} = \left(\mathbf{X}^-, \left(\mathbf{X}_e^0\right)_{e \in E_{int}(T)}, \mathbf{X}_1^+, \ldots, \mathbf{X}_d^+ \right) \,,$$

where

- $\mathbf{X}^- \in \mathfrak{X}_-^{\text{back}}(M, k(T))$ is written as $\mathbf{X}^- = \left(X_e^- \right)_{e \in E_{int}(T)}$,
- $\mathbf{X}_e^0 \in \mathfrak{X}_0^{\text{back}}(M, k(T) - 1)$ is denoted by

$$\mathbf{X}_e^0 = \left(\left(X_{ef}^0\right)_{f \in E_{int}(T) \setminus \{e\}}, X_{e+}, X_{e-} \right) \quad \text{for every } e \in E_{int}(T) \,,$$

- $\mathbf{X}_i^+ \in \mathfrak{X}_+^{\text{back}}(M, k(T))$ is given by

$$\mathbf{X}_i^+ = \left(X_e^+ \right)_{e \in E_{int}(T)} \quad \text{for every } i \in \{1, 2, \ldots, d\} \,.$$

Note that if $\mathbf{Y} \in \mathfrak{X}(T, \mathbf{X})$, then the above notation implies that

$$Y_0 \in \mathfrak{X}_-(M, k(T), \mathbf{X}^-) \,, \quad Y_e \in \mathfrak{X}_0(M, k(T), \mathbf{X}_e^0) \text{ and } Y_i \in \mathfrak{X}_+(M, k(T), \mathbf{X}_i^+)$$

for every $e \in E_{int}(T)$ and $i \in \{1, \ldots, d\}$.

Definition 4.14 Let $\mathbf{X} \in \mathfrak{X}^{\text{back}}(T)$ and $\mathbf{Y} \in \mathfrak{X}(T, \mathbf{X})$ and $x_0, x_1, \ldots, x_d \in \text{Crit } f$. *The moduli space of* $\mathbf{Y}$*-perturbed Morse ribbon trees modelled on* T *starting in* x_0 *and ending in* $x_1, \ldots, x_d$ *is defined as*

$$\mathcal{A}_\mathbf{Y}^d(x_0, x_1, \ldots, x_d, T)$$
$$:= \left\{ \left(\gamma_0, (l_e, \gamma_e)_{e \in E_{int}(T)}, \gamma_1, \ldots, \gamma_d\right) \mid \gamma_0 \in \mathcal{P}_-(x_0), \ \gamma_i \in \mathcal{P}_+(x_i) \ \forall i \in \{1, \ldots, d\} \,, \right.$$
$$l_e > 0, \ \gamma_e \in H^1([0, l_e], M) \ \forall e \in E_{int}(T) \,,$$
$$\dot{\gamma}_0(s) + (\nabla f)(\gamma_0(s)) + Y_0 \left((l_e)_{e \in E_{int}(T)}, s, \gamma_0(s)\right) = 0 \quad \forall s \in (-\infty, 0] \,,$$
$$\dot{\gamma}_i(s) + (\nabla f)(\gamma_i(s)) + Y_i \left((l_e)_{e \in E_{int}(T)}, s, \gamma_i(s)\right) = 0 \quad \forall s \in [0, +\infty) \ \forall i \in \{1, \ldots, d\} \,,$$
$$\dot{\gamma}_e(s) + (\nabla f)(\gamma_e(s)) + Y_e \left((l_f)_{f \in E_{int}(T) \setminus \{e\}}, l_e, s, \gamma_e(s)\right) = 0 \quad \forall s \in [0, l_e] \ \forall e \in E_{int}(T) \,,$$
$$\gamma_e(l_e) = \gamma_f(0) \text{ for all } e, f \in E_{int}(T) \text{ with } v_{\text{out}}(e) = v_{\text{in}}(f) \,,$$
$$\gamma_e(l_e) = \gamma_j(0) \text{ for all } e \in E_{int}(T), \ j \in \{1, \ldots, d\} \text{ with } v_{\text{out}}(e) = v_{\text{in}}(e_j(T)) \,,$$
$$\left. \gamma_0(0) = \gamma_e(0) \text{ for all } e \in E_{int}(T) \text{ with } v_{\text{out}}(e_0(T)) = v_{\text{in}}(e) \right\} \,.$$

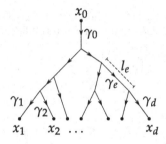

Fig. 4.3 An element of $\mathcal{A}_{\mathbf{Y}}^d(x_0, x_1, \ldots, x_d, T)$

See Fig. 4.3 for a picture of a perturbed Morse ribbon tree. The conditions in the last three lines of the definition of $\mathcal{A}_{\mathbf{Y}}^d(x_0, x_1, \ldots, x_d, T)$ can be loosely rephrased by saying that all the trajectories associated to the edges of T are coupled at the common associated vertices. They ensure that the family of trajectories induces a continuous map $T \to M$.

The proof of the following theorem is a major application of the results of Chap. 3 and particularly of Theorem 3.10.

Theorem 4.15 *Let* $\mathbf{X} \in \mathfrak{X}^{back}(T)$. *For generic choice of* $\mathbf{Y} \in \mathfrak{X}(T, \mathbf{X})$, *the moduli space* $\mathcal{A}_{\mathbf{Y}}^d(x_0, x_1, \ldots, x_d, T)$ *can be equipped with the structure of a manifold of class* C^{n+1} *of dimension*

$$\dim \mathcal{A}_{\mathbf{Y}}^d(x_0, x_1, \ldots, x_d, T) = \mu(x_0) - \sum_{j=1}^d \mu(x_j) + k(T)$$

for all $x_0, x_1, \ldots, x_d \in \mathrm{Crit}\ f$.

Proof In strict analogy with Remark 4.12, the endpoint conditions, i.e. the last three lines in the definition of $\mathcal{A}_{\mathbf{Y}}^d(x_0, x_1, \ldots, x_d, T)$, are equivalent to

$$\big(\gamma_0(0), (\gamma_e(0), \gamma_e(l_e))_{e \in E_{int}(T)}, \gamma_1(0), \ldots, \gamma_d(0)\big) \in \Delta_T .$$

Using the notation of Chap. 3, this condition can be reformulated as

$$\underline{E}_{\mathbf{Y}}\big(\gamma_0, (l_e, \gamma_e)_{e \in E_{int}(T)}, \gamma_1, \ldots, \gamma_d\big) \in \Delta_T ,$$

where $\underline{E}_{\mathbf{Y}}$ is defined as the corresponding product of evaluation maps as in the definition of the moduli space in Theorem 3.10.

This reformulation implies that in the notation of Theorem 3.10 the following equality holds true:

$$\mathcal{A}_{\mathbf{Y}}^d(x_0, x_1, \ldots, x_d, T) = \mathcal{M}_{\mathbf{Y}}\big(x_0, (x_1, \ldots, x_d), (1, k(T), d), (0, +\infty)^{k(T)}, \Delta_T\big) .$$

Thus, the statement is nothing but a straightforward consequence of Theorem 3.10. It remains to show that the claimed dimension formula coincides with the one from Theorem 3.10. The latter one yields:

$$\dim \mathcal{A}_{\mathbf{Y}}^d(x_0, x_1, \ldots, x_d, T, \mathbf{X}) = \mu(x_0) - \sum_{i=1}^{d} \mu(x_i) + (k(T) + d)n + k(T) - \operatorname{codim} \Delta_T$$

$$= \mu(x_0) - \sum_{i=1}^{d} \mu(x_i) + k(T),$$

where the second equality is a consequence of Proposition 4.11.   □

Definition 4.16 A perturbation $\mathbf{Y} \in \mathfrak{X}(T)$ for which $\mathcal{A}_{\mathbf{Y}}^d(x_0, x_1, \ldots, x_d, T, \mathbf{X})$ can be equipped with the structure of a manifold of class C^{n+1} for all $x_0, x_1, \ldots, x_d \in \operatorname{Crit} f$ is called *regular*.

Chapter 5
The Convergence Behaviour of Sequences of Perturbed Morse Ribbon Trees

Having constructed moduli spaces of perturbed Morse ribbon trees in the previous chapter, we want to investigate sequential compactness properties of these moduli spaces. Our starting point is the consideration of certain sequential compactness results for spaces of perturbed Morse trajectories of the three different types we introduced in Chap. 2. We will show that in all three cases, every sequence in the respective moduli space without a convergent subsequence has a subsequence that converges geometrically against a family of trajectories. The notion of geometric convergence will be made precise in Sects. 5.1 and 5.2.

Afterwards, we will draw conclusions from these results for spaces of perturbed Morse ribbon trees. The main result of this section is Theorem 5.10, following immediately from the compactness results for perturbed Morse trajectories. It roughly states that every sequence of perturbed Morse ribbon trees without a subsequence which converges in its respective domain, has a geometrically convergent subsequence. This notion of geometric convergence will be defined later in this chapter.

Having established the main compactness result, we will focus our attention on certain special cases. Theorems 5.15, 5.16 and 5.17 describe possibilities for the limiting behaviour of sequences of perturbed Morse ribbon trees in greater detail.

Finally, we formulate a theorem for describing the situation of simultaneous occurrence of the convergence phenomena described in the aforementioned three theorems. This theorem will enable us to investigate spaces of geometric limits of sequences of perturbed Morse ribbon trees. Eventually, we will use the structures of the limit spaces to construct higher order multiplications on Morse cochain complexes in Chap. 6.

The sequential compactness theorems for moduli spaces of perturbed Morse trajectories can be shown in strict analogy with the compactness results in the unperturbed case. The compactness result for semi-infinite Morse trajectories is shown in

© Springer International Publishing AG, part of Springer Nature 2018
S. Mescher, *Perturbed Gradient Flow Trees and A∞-algebra Structures in Morse Cohomology*, Atlantis Studies in Dynamical Systems 6,
https://doi.org/10.1007/978-3-319-76584-6_5

[Sch99, Sect. 4], while the compactness result for finite-length Morse trajectories is proven in [Weh12], along with a more general statement covering the semi-infinite case as well.

The line of argument of the respective reference can be transferred to the case of perturbed Morse trajectories with only minor changes. Therefore, we will omit parts of the proofs of the decisive compactness theorems in this chapter and directly apply the respective trajectory compactness theorem instead.

Throughout this chapter, we assume that every ribbon tree is equipped with an ordering of its internal edges.

All results of this chapter are independent of the chosen orderings. We will come back to these choices in Chap. 6 and Appendix A.

5.1   Geometric Convergence of Sequences of Semi-infinite Trajectories

We begin by stating the crucial sequential compactness theorem for perturbed semi-infinite Morse trajectories. For every $x, y \in \mathrm{Crit}\, f$ let

$$\widehat{\mathcal{M}}(x, y) := \widehat{\mathcal{M}}(x, y, g)$$

denote the space of unparametrized negative gradient flow lines of f with respect to g, as defined in the Chap. 1.

Theorem 5.1 *(i) Let $x \in \mathrm{Crit}\, f$, $\{\gamma_n\}_{n\in\mathbb{N}} \subset \mathcal{P}_-(x)$, $Y_\infty \in \mathfrak{X}_-(M)$ and $\{Y_n\}_{n\in\mathbb{N}}$ be a sequence in $\mathfrak{X}_-(M)$ which converges against Y_∞ and such that*

$$\gamma_n \in W^-(x, Y_n) \quad \text{for every } n \in \mathbb{N}.$$

If $\{\gamma_n\}_{n\in\mathbb{N}}$ does not have a subsequence which converges in $\mathcal{P}_-(x)$, then there will be $m \in \mathbb{N}$, $x_1, \ldots, x_m \in \mathrm{Crit}\, f$ with

$$\mu(x) > \mu(x_1) > \cdots > \mu(x_m)$$

and curves

$$(\hat{g}_1, \ldots, \hat{g}_m, \gamma_-) \in \widehat{\mathcal{M}}(x, x_1) \times \widehat{\mathcal{M}}(x_1, x_2) \times \cdots \times \widehat{\mathcal{M}}(x_{m-1}, x_m) \times W^-(x_m, Y_\infty),$$

as well as a sequence $\{\tau_q^j\}_{q\in\mathbb{N}} \subset \mathbb{R}$ diverging to $-\infty$ for every $j \in \{1, 2, \ldots, m\}$ and a subsequence $\{\gamma_{n_q}\}_{q\in\mathbb{N}}$ such that

- *$\{\gamma_{n_q}\}_{q\in\mathbb{N}}$ converges against γ_- in the C^∞_{loc}-topology,*

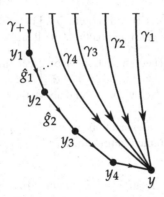

Fig. 5.1 A geometrically convergent sequence of perturbed semi-infinite half-trajectories ending in $y \in \mathrm{Crit}\, f$

- $\left\{ \gamma_{n_q} \left(\cdot + \tau_q^j \right) \Big|_{(-\infty, T]} \right\}_{q \geq q_{T,j}}$ *converges against* $g_j|_{(-\infty, T]}$ *in the* C_{loc}^∞-*topology for all* $T > 0$ *and* $q_{T,j} \in \mathbb{N}$ *for which the restrictions are well-defined, where* g_j *is a representative of* $\hat{g}_j$ *for every* $j \in \{1, 2, \ldots, m\}$.

(ii) *Let* $y \in \mathrm{Crit}\, f$, $\{\gamma_n\}_{n \in \mathbb{N}} \subset \mathcal{P}^+(y)$, $\{Y_n\}_{n \in \mathbb{N}} \subset \mathfrak{X}_+(M)$ *and* $Y_\infty \in \mathfrak{X}_+(M)$ *such that* $\{Y_n\}_{n \in \mathbb{N}}$ *converges against* Y_∞ *and*

$$\gamma_n \in W^+(x, Y_n) \quad \text{for every } n \in \mathbb{N}.$$

If $\{\gamma_n\}_{n \in \mathbb{N}}$ *does not have a subsequence which converges in* $\mathcal{P}_+(y)$, *then there will be* $m \in \mathbb{N}$, $y_1, \ldots, y_m \in \mathrm{Crit}\, f$ *with*

$$\mu(y_1) > \cdots > \mu(y_m) > \mu(y)$$

and curves

$$(\gamma_+, \hat{g}_1, \ldots, \hat{g}_m) \in W^+(y_1, Y_\infty) \times \widehat{\mathcal{M}}(y_1, y_2) \times \cdots \times \widehat{\mathcal{M}}(y_{m-1}, y_m) \times \widehat{\mathcal{M}}(y_m, y),$$

as well a sequence $\{\tau_q^j\}_{q \in \mathbb{N}} \subset \mathbb{R}$ *diverging to* $+\infty$ *for every* $j \in \{1, \ldots, m\}$ *and a subsequence* $\{\gamma_{n_q}\}_{q \in \mathbb{N}}$ *such that*

- $\{\gamma_{n_q}\}_{q \in \mathbb{N}}$ *converges against* γ_+ *in the* C_{loc}^∞-*topology,*
- $\left\{ \gamma_{n_q} \left(\cdot + \tau_q^j \right) \Big|_{[-T, +\infty)} \right\}_{q \geq q_{T,j}}$ *converges against* $g_j|_{(-T, +\infty)}$ *in the* C_{loc}^∞-*topology for every* $T > 0$ *and every* $q_{n,T} \in \mathbb{N}$ *such that the restrictions are well-defined, for a representative* g_j *of* $\hat{g}_j$ *for every* $j \in \{1, \ldots, m\}$.

Theorem 5.1 is shown by applying the methods used to prove the corresponding result for unperturbed half-trajectories. For compactness theorems for spaces of unperturbed Morse trajectories, see [Weh12, Theorem 2.6] or [Sch93, Sect. 2.4] for analogous results for trajectories defined on $\mathbb{R}$.

Figure 5.1 illustrates part (ii) of Theorem 5.1. For an illustration of part (i), turn Fig. 5.1 upside down and invert the directions of all arrows.

Remark 5.2 Concerning Theorem 5.1, it is of great importance that the spaces $\mathfrak{X}_\pm(M)$ are defined such that for every $Y \in \mathfrak{X}_\pm(M)$, the value $Y(t, x)$ vanishes whenever t lies below the fixed value -1 and above the fixed value 1 in the positive case.

If the perturbations were allowed to be non-vanishing for arbitrary time parameters, the sequences of reparametrization times $\{\tau_n^j\}_{n \in \mathbb{N}}$ would in general not exist in the above setting.

Definition 5.3 (i) In the situation of Theorem 5.1 (i), we say that $\{\gamma_n\}_{n \in \mathbb{N}}$ *converges geometrically against* $(\hat{g}_1, \ldots, \hat{g}_m, \gamma_-)$ and we call $(\hat{g}_1, \ldots, \hat{g}_m, \gamma_-)$ the *geometric limit of* $\{\gamma_n\}_{n \in \mathbb{N}}$.

(ii) In the situation of Theorem 5.1 (ii), we say that $\{\gamma_n\}_{n \in \mathbb{N}}$ *converges geometrically against* $(\gamma_+, \hat{g}_1, \ldots, \hat{g}_m)$ and we call $(\gamma_+, \hat{g}_1, \ldots, \hat{g}_m)$ the *geometric limit of* $\{\gamma_n\}_{n \in \mathbb{N}}$.

Remark 5.4 The following special case of Theorem 5.1 will be relevant in the discussion of the convergence of perturbed Morse ribbon trees. Let $Y \in \mathfrak{X}_-(M, k, \mathbf{X})$ for some $\mathbf{X} = (X_1, \ldots, X_k) \in \mathfrak{X}_-^{\text{back}}(M, k)$. Let $\{(l_{1n}, \ldots, l_{kn})\}_{n \in \mathbb{N}}$ be a sequence in $(0, +\infty)^k$, such that $\{l_{in}\}_{n \in \mathbb{N}}$ diverges against $+\infty$ for a unique $i \in \{1, 2, \ldots, k\}$ and that $\{l_{jn}\}_{n \in \mathbb{N}}$ converges against a number $l_{j\infty} \in [0, +\infty)$ if $j \neq i$.

Define $Y_n \in \mathfrak{X}_-(M)$ for every $n \in \mathbb{N}$ by

$$Y_n(t, x) := Y((l_{1n}, \ldots, l_{kn}), t, x) \quad \forall t \in (-\infty, 0], \quad x \in M,$$

let $x \in \text{Crit } f$ and let $\{\gamma_n\}_{n \in \mathbb{N}}$ be a sequence with $\gamma_n \in W^-(x, Y_n)$ for every $n \in \mathbb{N}$.

If $\{\gamma_n\}_{n \in \mathbb{N}}$ converges geometrically in this situation, then the it will hold for the geometric limit $(\hat{g}_1, \ldots, \hat{g}_m, \gamma_-)$ by definition of a background perturbation that

$$\gamma_- \in W^-(x_m, X_i(l_{1\infty}, \ldots, l_{(i-1)\infty}, l_{(i+1)\infty}, \ldots, l_{k\infty})) .$$

The analogous remark is true for positive semi-infinite half-trajectories.

5.2 Geometric Convergence of Sequences of Finite-Length Trajectories

The convergence theorem for perturbed finite-length Morse trajectories takes a similar form, but we need to introduce the right notion of convergence in advance.

Definition 5.5 Let $k \in \mathbb{N}_0$, let $\{Y_n\}_{n \in \mathbb{N}} \subset \mathfrak{X}_0(M)$ be a sequence converging against $Y_\infty \in \mathfrak{X}_0(M)$ and let $\{(l_n, \gamma_n)\}_{n \in \mathbb{N}}$ be a sequence, where $l_n \geq 0$, $\gamma_n : [0, l_n] \to M$ and

$$(l_n, \gamma_n) \in \mathcal{M}(Y_n)$$

for every $n \in \mathbb{N}$. The sequence $\{(l_n, \gamma_n)\}_{n \in \mathbb{N}}$ is called *convergent* iff the sequence

$$\{\psi_{Y_n}(l_n, \gamma_n)\}_{n \in \mathbb{N}} \subset \mathcal{M}(f, g)$$

converges in $\mathcal{M}(f, g)$. Here, we use the map ψ from (2.15) and put $\psi_{Y_n} := \psi(Y_n, \cdot)$ for every $n \in \mathbb{N}$. Every such

$$\psi_{Y_n} : \mathcal{M}(Y_n) \xrightarrow{\cong} \mathcal{M}(f, g)$$

is a diffeomorphism of class C^{n+1}.

Theorem 5.6 *Let $k \in \mathbb{N}_0$ and let $\{Y_n\}_{n \in \mathbb{N}} \subset \mathfrak{X}_0(M)$ be a sequence converging against some $Y_\infty \in \mathfrak{X}_0(M)$. Consider a sequence $\{(l_n, \gamma_n)\}_{n \in \mathbb{N}}$ with*

$$(l_n, \gamma_n) \in \mathcal{M}(Y_n)$$

for every $n \in \mathbb{N}$. If the sequence $\{(l_n, \gamma_n)\}_{n \in \mathbb{N}}$ does not have a convergent subsequence, then there are a subsequence $\{(l_{n_q}, \gamma_{n_q})\}_{q \in \mathbb{N}}$ with

$$\lim_{q \to \infty} l_{n_q} = +\infty ,$$

$m \in \mathbb{N}$, $x_1, \ldots, x_m \in \mathrm{Crit}\, f$, $\{s_q^j\}_{q \in \mathbb{N}} \subset \mathbb{R}$ for every $j \in \{1, \ldots, m-1\}$ as well as curves

$$\left(\gamma_+, \hat{g}_1, \ldots, \hat{g}_{m-1}, \gamma_-\right) \in W^+(x_1, Y_+) \times \widehat{\mathcal{M}}(x_1, x_2) \times \ldots \times \widehat{\mathcal{M}}(x_{m-1}, x_m)$$
$$\times W^-(x_m, Y_-)$$

where $Y_+ \in \mathfrak{X}_+(M)$ and $Y_- \in \mathfrak{X}_-(M, k)$ are defined by

$$(Y_+, Y_-) = \lim_{q \to \infty} \mathrm{split}_0\left(l_{n_q}, Y_{n_q}\right) .$$

with split_0 is given as in (2.9), such that:

- $\left\{\gamma_{n_q}|_{[0,T]}\right\}_{q \geq q_T}$ *converges against $\gamma_+|_{[0,T]}$ in the C^∞-topology for every $T > 0$ and $q_T \in \mathbb{N}$ such that the restrictions are well-defined,*
- $\left\{\gamma_{n_q}\left(\cdot + s_q^j\right)\Big|_{[-T,T]}\right\}_{q \geq q_{T,j}}$ *converges against $g_j|_{[-T,T]}$ in the C^∞-topology for all $T > 0$ and $q_{T,j} \in \mathbb{N}$ for which the restrictions are well-defined, where g_j is a representative of $\hat{g}_j$ for every $j \in \{1, \ldots, m-1\}$,*
- $\left\{\gamma_{n_q}\left(\cdot + l_{n_q}\right)\Big|_{[-T,0]}\right\}_{q \geq q_T}$ *converges against $\gamma_-|_{[-T,0]}$ in the C^∞-topology for all $T > 0$ and $q_T \in \mathbb{N}$ for which the restrictions are well-defined.*

Remark 5.7 For the validity of Theorem 5.6, the compactness of M is required. If M was non-compact, there might be sequences of finite-length trajectories whose interval lengths tends to infinity, but which are not geometrically convergent.

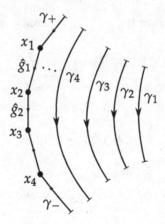

Fig. 5.2 A geometrically convergent sequence of perturbed finite-length Morse trajectories

Speaking in terms of geometric intuition this corresponds to the starting points or end points of the trajectory sequence "escaping to infinity". This phenomenon can obviously not occur in compact manifolds.

Definition 5.8 In the situation of Theorem 5.6, we say that $\{(l_{n_q}, \gamma_{n_q})\}_{q \in \mathbb{N}}$ *converges geometrically against* $(\gamma_+, \hat{g}_1, \ldots, \hat{g}_{m-1}, \gamma_-)$ and we call $(\gamma_+, \hat{g}_1, \ldots, \hat{g}_{m-1}, \gamma_-)$ the *geometric limit of* $\{(l_{n_q}, \gamma_{n_q})\}_{q \in \mathbb{N}}$.

See Fig. 5.2 for an illustration of the geometric convergence of Theorem 5.6.

5.3 A First Sequential Compactness Result for Perturbed Morse Ribbon Trees

We want to derive a convergence theorem for sequences of perturbed Morse ribbon trees from Theorems 5.1 and 5.6. Before stating the theorem, we will make some general observations and then motivate it by drawing immediate conclusions from Theorems 5.1 and 5.6.

We consider a sequence of perturbed Morse ribbon trees

$$\left\{\underline{\gamma}_n\right\}_{n \in \mathbb{N}} = \left\{\left(\gamma_{0n}, (l_{en}, \gamma_{en})_{n \in \mathbb{N}}, \gamma_{1n}, \ldots, \gamma_{dn}\right)\right\}_{n \in \mathbb{N}} \subset \mathcal{A}_{\mathbf{Y}}^d(x_0, x_1, \ldots, x_d, T)$$

for some $d \geq 2$, $T \in \mathrm{RTree}_d$, $x_0, x_1, \ldots, x_d \in \mathrm{Crit}\, f$ and

$$\mathbf{Y} = (Y_0, (Y_e)_{e \in E_{int}(T)}, Y_1, \ldots, Y_d) \in \mathfrak{X}(T) \,.$$

By definition of $\mathcal{A}_{\mathbf{Y}}^d(x_0, x_1, \ldots, x_d, T)$, we can regard each component sequence of $\left\{\underline{\gamma}_n\right\}_n$ as a sequence of perturbed Morse trajectories. Before applying the compactness results from above to the component sequences, we make an observation on the convergence of the edge length sequences.

The following lemma is a simple consequence of the Bolzano–Weierstraß theorem.

Lemma 5.9 *Let $k \in \mathbb{N}$ and $\left\{ \vec{l}_n = (l_{1n}, \ldots, l_{kn}) \right\}_{n \in \mathbb{N}} \subset [0, +\infty)^k$ be a sequence. Then there are a subsequence $\left\{ \vec{l}_{n_q} \right\}_{q \in \mathbb{N}}$ and two disjoint subsets*

$$I_1, I_2 \subset \{1, 2, \ldots, k\} \quad \text{with} \quad I_1 \cup I_2 = \{1, 2, \ldots, k\}, \tag{5.1}$$

such that:

- *for every $i \in I_1$, the sequence $\{l_{in_q}\}_{q \in \mathbb{N}} \subset [0, +\infty)$ diverges to $+\infty$,*
- *for every $i \in I_2$, the sequence $\{l_{in_q}\}_{q \in \mathbb{N}}$ converges with $l_{i\infty} := \lim_{q \to \infty} l_{in_q}$.*

In the situation of Lemma 5.9, define $\vec{l}_\infty \in [0, +\infty)^{|I_2|}$ by

$$\vec{l}_\infty := (l_{i\infty})_{i \in I_2}. \tag{5.2}$$

We want to apply Lemma 5.9 to the sequence $\left\{ \underset{n}{\gamma} \right\}_{n \in \mathbb{N}}$ from above. For all $n \in \mathbb{N}$ we let $\vec{l}_n \in [0, +\infty)^{k(T)}$ denote the vector whose entries are given by the family $(l_{en})_{e \in E_{int}(T)}$, ordered by the given ordering of $E_{int}(T)$. We assume w.l.o.g. that every component sequence of $\left\{ \vec{l}_n \right\}_{n \in \mathbb{N}} \subset (0, +\infty)^{k(T)}$ either converges or diverges to $+\infty$, since by Lemma 5.9 this property holds true up to passing to a subsequence.

Then the following will hold for every $n \in \mathbb{N}$:

$$\gamma_{0n} \in W^- \left(x_0, Y_0 \left(\vec{l}_n \right) \right), \quad \gamma_{in} \in W^+ \left(x_i, Y_i \left(\vec{l}_n \right) \right) \ \forall i \in \{1, 2, \ldots, d\},$$

$$(l_{en}, \gamma_{en}) \in \mathcal{M} \left(Y_e \left((l_{fn})_{f \in E_{int}(T) \setminus \{e\}} \right) \right) \quad \forall e \in E_{int}(T).$$

By applying Theorem 5.1 with $Y_n = Y_i \left(\vec{l}_n \right)$ for every $n \in \mathbb{N}$ we derive that at least one of the following is true for each $i \in \{0, 1, \ldots, d\}$, i.e. for each of the component sequences associated with external edges of T:

- $\{\gamma_{in}\}_{n \in \mathbb{N}}$ has a subsequence that converges in $\mathcal{P}_\pm(x_i)$,
- $\{\gamma_{in}\}_{n \in \mathbb{N}}$ has a subsequence that converges geometrically against a family of trajectories.

Applying Theorem 5.6 with $Y_n = Y_e((l_{fn})_{f \in E_{int}(T) \setminus \{e\}})$ to the component sequences associated with elements of $E_{int}(T)$ yields that for every $e \in E_{int}(T)$ at least one of the following holds:

- $\{(l_{en}, \gamma_{en})\}_{n \in \mathbb{N}}$ has a subsequence that converges in the sense of Definition 5.5 and the length of its limit curve is positive,

- $\{(l_{en}, \gamma_{en})\}_{n\in\mathbb{N}}$ has a subsequence that converges in the sense of Definition 5.5 and the length of its limit curve is zero,
- $\{(l_{en}, \gamma_{en})\}_{n\in\mathbb{N}}$ has a subsequence that converges geometrically against a family of trajectories.

In both situations, we have used that by definition of the spaces $\mathfrak{X}_\pm(M, k)$ and $\mathfrak{X}_0(M, k)$, the respective sequence $\{Y_n\}_{n\in\mathbb{N}}$ from above is indeed convergent in the respective space.

Moreover, since $\left\{\underline{\gamma}_n\right\}_{n\in\mathbb{N}}$ has finitely many components, we can find a "common" subsequence for all edges of T such that one of the above holds. More precisely, there is a subsequence $\left\{\underline{\gamma}_{n_q}\right\}_{q\in\mathbb{N}}$ such that all the sequences $\{\gamma_{in_q}\}_{q\in\mathbb{N}}$, $i \in \{0, 1, \dots, d\}$, and $\{(l_{en_q}, \gamma_{en_q})\}_{q\in\mathbb{N}}$, $e \in E_{int}(T)$, either converge or converge geometrically.

While we have not distinguished between the first two bullet points for finite-length components in Theorem 5.6, we will do so in the following considerations. The reason for this distinction lies in the definition of the moduli spaces of perturbed Morse ribbon trees.

Suppose that every component sequence of $\left\{\underline{\gamma}_n\right\}_{n\in\mathbb{N}}$ converges. If $\lim_{n\to\infty} l_{en} > 0$ for every $e \in E_{int}(T)$, then the family of the limits will again be identified with an element of $\mathcal{A}_Y^d(x_0, x_1, \dots, x_d, T)$.

If there is an $e \in E_{int}(T)$ for with $\lim_{n\to\infty} l_{en} = 0$, then there will be no such identification, since by definition of $\mathcal{A}_Y^d(x_0, x_1, \dots, x_d, T)$, every finite-length component of an element of this moduli space must have positive length. Instead, in the course of this section we will identify such limits with perturbed Morse ribbon trees *modelled on different trees than* T.

These considerations motivate the distinction between the sets E_1 and E_3 in the following convergence theorem for sequences of perturbed Morse ribbon trees, which summarizes our elaborations on components of sequences of perturbed Morse ribbon trees.

Theorem 5.10 *Let* $d \geq 2$, $T \in \mathrm{RTree}_d$, $\mathbf{Y} \in \mathfrak{X}(T)$ *and* $x_0, x_1, \dots, x_d \in \mathrm{Crit} f$. *Let*

$$\left\{\left(\gamma_{0n}, (l_{en}, \gamma_{en})_{n\in\mathbb{N}}, \gamma_{1n}, \dots, \gamma_{dn}\right)\right\}_{n\in\mathbb{N}}$$

be a sequence in $\mathcal{A}_Y^d(x_0, x_1, \dots, x_d, T)$. *For every* $e \in E(T)$ *define a sequence* $\{\bar{\gamma}_{en}\}_{n\in\mathbb{N}}$ *by putting*

$$\bar{\gamma}_{en} := \begin{cases} \gamma_{in} & \text{if } e = e_i(T) \text{ for some } i \in \{0, 1, \dots, d\}, \\ (l_{en}, \gamma_{en}) & \text{if } e \in E_{int}(T), \end{cases}$$

for every $n \in \mathbb{N}$. *There are sets* $E_1, E_2 \subset E(T)$, $E_3 \subset E_{int}(T)$ *with*

$$E(T) = E_1 \sqcup E_2 \sqcup E_3,$$

such that

- *for every* $e \in E_1$, *the sequence* $\{\bar{\gamma}_{en}\}_{n\in\mathbb{N}}$ *has a convergent subsequence and*

$$\liminf_{n\to\infty} l_{en} > 0 \quad \text{if } e \in E_{int}(T) \, ,$$

- *for every* $e \in E_2$, *the sequence* $\{\bar{\gamma}_{en}\}_{n\in\mathbb{N}}$ *has a geometrically convergent subsequence,*
- *for every* $e \in E_3$, *the sequence* $\{(l_{en}, \gamma_{en})\}_{n\in\mathbb{N}}$ *has a convergent subsequence*

$$\{(l_{en_q}, \gamma_{en_q})\}_{q\in\mathbb{N}} \quad \text{with} \quad \lim_{q\to\infty} l_{en_q} = 0 \, .$$

Theorem 5.10 immediately leads to the following notion of convergence for sequences of perturbed Morse ribbon trees:

Definition 5.11 Let $\left\{\left(\gamma_{0n}, (l_{en}, \gamma_{en})_{e\in E_{int}(T)}, \gamma_{1n}, \ldots, \gamma_{dn}\right)\right\}_{n\in\mathbb{N}}$ be a sequence of perturbed Morse ribbon trees.

1. Let $i \in \{0, 1, \ldots, d\}$. Whenever the sequence $\{\gamma_{in}\}_{n\in\mathbb{N}}$ converges, we denote its limit by $\gamma_{i\infty}$.
2. Let $e \in E_{int}(T)$. Whenever the sequence $\{(l_{en}, \gamma_{en})\}_{n\in\mathbb{N}}$ converges, we denote its limit by $(l_{e\infty}, \gamma_{e\infty})$.
3. We say that the sequence is *convergent* if every component sequence $\{\bar{\gamma}_{en}\}_{n\in\mathbb{N}}$, $e \in E(T)$, converges (where we have used the notation from Theorem 5.10).
4. The *limit* of a convergent sequence of perturbed Morse ribbon trees is defined as the product of the limits of the component sequences.

Note that a sequence of perturbed Morse ribbon trees has a convergent subsequence in the sense of the previous definition if and only if we can choose $E_2 = \emptyset$ in Theorem 5.10.

5.4 Convergent Subsequences and Edge Collapsing

In this and the following two sections we will describe the convergence behaviour of sequences of perturbed Morse ribbon trees in several special cases of Theorem 5.10 in greater detail. More precisely, we will

- consider the case $E_1 = E(T) \setminus F$, $E_2 = \emptyset$ and $E_3 = F$ for some $F \subset E_{int}(T)$ in this section,
- consider the case $E_1 = E(T) \setminus \{e_i(T)\}$, $E_2 = \{e_i(T)\}$ and $E_3 = \emptyset$ for some $i \in \{0, 1, \ldots, d\}$ in Sect. 5.6,
- consider the case $E_1 = E(T) \setminus E_2$, $E_2 \subset E_{int}(T)$ and $E_3 = \emptyset$ in Sect. 5.6.

After these investigations, we return to the general case combining the results of the three cases.

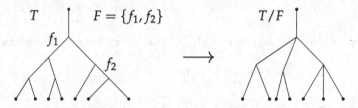

Fig. 5.3 An example of the collapse of internal edges of a ribbon tree

Definition 5.12 Let $d \in \mathbb{N}$, $d \geq 2$. For $T \in \mathrm{RTree}_d$ and $e \in E_{int}(T)$, we define

$$T/e \in \mathrm{RTree}_d$$

as the unique tree we obtain from T after collapsing the edge e. More precisely, T/e is the unique tree with

$$E(T/e) = E(T) \setminus \{e\}$$

and such that $v_{\mathrm{out}}(f) = v_{\mathrm{in}}(f')$ for every $f, f' \in E(T) \setminus \{e\}$ with

$$v_{\mathrm{out}}(f) = v_{\mathrm{in}}(e) \quad \text{and} \quad v_{\mathrm{in}}(f') = v_{\mathrm{out}}(e)\,.$$

Let $d \in \mathbb{N}$, $d \geq 2$. For $T \in \mathrm{RTree}_d$ and $F \subset E_{int}(T)$, we define

$$T/F \in \mathrm{RTree}_d$$

as the unique tree we obtain from T after collapsing every edge contained in F. More precisely, if $F = \{f_1, f_2, \ldots, f_m\}$ for suitable $m \in \mathbb{N}$, then we define T/F inductively by

$$T/F := (\ldots(((T/f_1)/f_2)/f_3)\ldots)/f_m\,.$$

One checks that the tree T/F is independent of the choice of ordering of F.

Figure 5.3 shows an example of such a "quotient tree". Note that especially

$$k(T/F) = k(T) - |F|$$

for all $T \in \bigcup_{d \geq 2} \mathrm{RTree}_d$ and $F \subset E_{int}(T)$.

In the following we will relate moduli spaces of perturbed Morse ribbon trees modelled on T with those modelled on T/F for some $F \subset E_{int}(T)$. For this purpose, we need to find a way of considering perturbations in $\mathfrak{X}(T)$ and $\mathfrak{X}(T/F)$ at the same time. Before we investigate these relations, we therefore introduce a technique for relating perturbations associated with different ribbon trees to each other.

Definition 5.13 Let $d \in \mathbb{N}$, $T \in \mathrm{RTree}_d$ and $F \subset E_{int}(T)$.

1. For $Y \in \mathfrak{X}_{\pm}(M, k(T))$ we define

$$Y/F \in \mathfrak{X}_{\pm}(M, k(T/F)), \quad (Y/F)\left((l_e)_{e \in E_{int}(T/F)}, t, x\right) := Y\left((\tilde{l}_e)_{e \in E_{int}(T)}, t, x\right) ,$$

where $\tilde{l}_e := \begin{cases} l_e & \text{if } e \in E_{int}(T) \setminus F , \\ 0 & \text{if } e \in F . \end{cases}$

Analogously, for $Z \in \mathfrak{X}_0(M, k(T) - 1)$ and $f \in E_{int}(T) \setminus F$ we define

$$Z/F \in \mathfrak{X}_0(M, k(T/F) - 1),$$

$$(Z/F)\left((l_e)_{e \in E_{int}(T/F) \setminus \{f\}}, l_f, t, x\right) := Z\left((\tilde{l}_e)_{e \in E_{int}(T) \setminus \{f\}}, l_f, t, x\right) ,$$

where $\tilde{l}_e$ is given as above.

2. We define a map

$$\pi_F : \mathfrak{X}(T) \to \mathfrak{X}(T/F) , \tag{5.3}$$

$$\left(Y_0, (Y_e)_{e \in E_{int}(T)}, Y_1, \ldots, Y_d\right) \mapsto \left(Y_0/F, (Y_e/F)_{e \in E_{int}(T) \setminus F}, Y_1/F, \ldots, Y_d/F\right) ,$$

π_F is a composition of an evaluation map and a projection. Therefore, π_F is easily seen to be continuous and surjective.

3. Assume that $k := k(T) > 0$. We define maps

$$\pi_{\pm,F}^{back} : \mathfrak{X}_{\pm}^{back}(M, k) \to \mathfrak{X}_{\pm}^{back}(M, k(T/F)) , \quad (X_1, \ldots, X_k) \mapsto (X_1/F, \ldots, X_k/F) .$$

If $k(T/F) > 0$, then we further let $e \in E_{int}(T) \setminus F$ and define

$$\pi_{0,F}^{back} : \mathfrak{X}_0^{back}(M, k(T) - 1) \to \mathfrak{X}_0^{back}(M, k(T/F) - 1) ,$$

$$((X_f)_{f \in E_{int}(T) \setminus \{e\}}, X_+, X_-) \mapsto ((X_f/F)_{f \in E_{int}(T/F) \setminus \{e\}}, X_+/F, X_-/F) .$$

In terms of these maps, we define for $k(T/F) > 0$ a map

$$\pi_F^{back} : \mathfrak{X}^{back}(T) \to \mathfrak{X}^{back}(T/F) ,$$

$$(\mathbf{X}_0, (\mathbf{X}_e)_e, \mathbf{X}_1, \ldots, \mathbf{X}_d) \mapsto \left(\pi_{-,F}^{back}(\mathbf{X}_0), (\pi_{0,F}^{back}(\mathbf{X}_e))_e, \pi_{+,F}^{back}(\mathbf{X}_1), \ldots, \pi_{+,F}^{back}(\mathbf{X}_d)\right) ,$$

and for $k(T/F) = 0$ a map

$$\pi_F^{back} : \mathfrak{X}^{back}(T) \to \mathfrak{X}_-(M) \times \mathfrak{X}_+(M)^d ,$$

$$(\mathbf{X}_0, (\mathbf{X}_e)_e, \mathbf{X}_1, \ldots, \mathbf{X}_d) \mapsto \left(\pi_{-,F}^{back}(\mathbf{X}_0), \pi_{+,F}^{back}(\mathbf{X}_1), \ldots, \pi_{+,F}^{back}(\mathbf{X}_d)\right) ,$$

In analogy with π_F, the map π_F^{back} is continuous and surjective.

To relate perturbations in $\mathfrak{X}(T)$ with those in $\mathfrak{X}(T/F)$, we further need to develop a formalism for simultaneous choices of perturbations for families of ribbon trees. At the same time, we introduce analogous notions for background perturbations, which will not be required in this section, but in the following one.

Definition 5.14   1. A *d-perturbation datum* is a family $\mathbf{Y} = (\mathbf{Y}_T)_{T \in \mathrm{RTree}_d}$, such that $\mathbf{Y}_T \in \mathfrak{X}(T)$ for each $T \in \mathrm{RTree}_d$.

2. A *d*-perturbation datum $\mathbf{Y} = (\mathbf{Y}_T)_{T \in \mathrm{RTree}_d}$ is called *universal* if the following holds for all $T \in \mathrm{RTree}_d$ and $F \subset E_{int}(T)$:

$$\pi_F(\mathbf{Y}_T) = \mathbf{Y}_{T/F} .$$

3. Given a *d*-perturbation datum $\mathbf{Y} = (\mathbf{Y}_T)_{T \in \mathrm{RTree}_d}$ and for $T \in \mathrm{RTree}_d$ we write

$$\mathcal{A}_{\mathbf{Y}}^d(x_0, x_1, \ldots, x_d, T) := \mathcal{A}_{\mathbf{Y}_T}^d(x_0, x_1, \ldots, x_d, T)$$

   for all $x_0, x_1, \ldots, x_d \in \mathrm{Crit}\, f$.

4. Let $\mathrm{RTree}_d^* := \{T \in \mathrm{RTree}_d \mid k(T) > 0\}$. A *background d-perturbation datum* is a family $\mathbf{X} = (\mathbf{X}_T)_{T \in \mathrm{RTree}_d^*}$, such that $\mathbf{X}_T \in \mathfrak{X}^{\mathrm{back}}(T)$ for each $T \in \mathrm{RTree}_d^*$.

5. We call a background *d*-perturbation datum $\mathbf{X} = (\mathbf{X}_T)_{T \in \mathrm{RTree}_d^*}$ *universal* if is satisfies both of the following conditions:

   - for all $T \in \mathrm{RTree}_d^*$ and $F \subset E_{int}(T)$ with $k(T/F) > 0$ it holds that

   $$\pi_F^{\mathrm{back}}(\mathbf{X}_T) = \mathbf{X}_{T/F} .$$

   - for all $T, T' \in \mathrm{RTree}_d^*$ it holds with $F := E_{int}(T)$ and $F' := E_{int}(T')$ that

   $$\pi_F^{\mathrm{back}}(\mathbf{X}_T) = \pi_{F'}^{\mathrm{back}}(\mathbf{X}_{T'}) .$$

The next theorem describes the case that all component sequences of a sequence of perturbed Morse ribbon trees converge and that there are internal edges whose associated sequences of edge lengths tend to zero. This corresponds to the case

$$E_1 = E(T) \setminus F , \quad E_2 = \emptyset , \quad E_3 = F$$

for some $F \subset E_{int}(T)$ in Theorem 5.10.

Theorem 5.15 *Let* $d \in \mathbb{N}$, $d \geq 2$, $T \in \mathrm{RTree}_d$ *and let* $\mathbf{Y} \in \prod_{T \in \mathrm{RTree}_d} \mathfrak{X}(T)$ *be a universal d-perturbation datum. Let* $x_0, x_1, \ldots, x_d \in \mathrm{Crit}\, f$ *and let*

$$\left\{ \left(\gamma_{0n}, (l_{en}, \gamma_{en})_{e \in E_{int}(T)}, \gamma_{1n}, \ldots, \gamma_{dn} \right) \right\}_{n \in \mathbb{N}} \subset \mathcal{A}_{\mathbf{Y}}^d(x_0, x_1, \ldots, x_d, T)$$

be a sequence, for which all of the sequences $\{\gamma_{in}\}_{n \in \mathbb{N}}$, $i \in \{0, 1, \ldots, d\}$, *and* $\{(l_{en}, \gamma_{en})\}_{n \in \mathbb{N}}$, $e \in E_{int}(T)$, *converge and for which there is an* $F \subset E_{int}(T)$ *such that*

$$l_{f\infty} = 0 \text{ for every } f \in F \,,$$
$$l_{f\infty} > 0 \text{ for every } f \in E_{int}(T) \setminus F \,.$$

Then $\left(\gamma_{0\infty}, (l_{e\infty}, \gamma_{e\infty})_{e \in E_{int}(T)\setminus F}, \gamma_{1\infty}, \ldots, \gamma_{d\infty}\right) \in \mathcal{A}_Y^d(x_0, x_1, \ldots, x_d, T/F)$.

Proof By definition of $\mathcal{A}_Y^d(x_0, x_1, \ldots, x_d, T)$, it holds for every $n \in \mathbb{N}$ that

$$\left(\gamma_{0n}(0), (\gamma_{en}(0), \gamma_{en}(l_{en}))_{e \in E_{int}(T)}, \gamma_{1n}(0), \ldots, \gamma_{dn}(0)\right) \in \Delta_T \,.$$

Since Δ_T is closed, the convergence of the sequence implies

$$\lim_{n \to \infty} \left(\gamma_{0n}(0), (\gamma_{en}(0), \gamma_{en}(l_{en}))_{e \in E_{int}(T)}, \gamma_{1n}(0), \ldots, \gamma_{dn}(0)\right) \in \Delta_T \,.$$

Moreover, since $\lim_{n \to \infty} l_{fn} = 0$ for every $f \in F$, we conclude

$$\lim_{n \to \infty} \gamma_{fn}(0) = \lim_{n \to \infty} \gamma_{fn}(l_{fn}) \quad \forall f \in F \,,$$

which yields

$$\lim_{n \to \infty} \left(\gamma_{0n}(0), (\gamma_{en}(0), \gamma_{en}(l_{en}))_{e \in E_{int}(T)}, \gamma_{1n}(0), \ldots, \gamma_{dn}(0)\right)$$
$$\in \left\{ (q_0, (q_{in}^e, q_{out}^e)_{e \in E_{int}(T)}, q_1, \ldots, q_d) \in \Delta_T \,\middle|\, q_{in}^f = q_{out}^f \ \forall f \in F \right\} \,. \quad (5.4)$$

From the definition of the T-diagonal (see Remark 4.12) we derive

$$\left\{ (q_0, (q_{in}^e, q_{out}^e)_{e \in E_{int}(T)}, q_1, \ldots, q_d) \in \Delta_T \,\middle|\, q_{in}^f = q_{out}^f \ \forall f \in F \right\}$$
$$= \left\{ (q_0, (q_{in}^e, q_{out}^e)_{e \in E_{int}(T)}, q_1, \ldots, q_d) \in \Delta_T \,\middle|\, q_{in}^e = q_{out}^{e'} \ \forall e, e' \in E_{int}(T) \text{ s.t. }\right.$$
$$\left. \exists f_1, \ldots, f_m \in F \text{ with } q_{out}^e = q_{in}^{f_1}, \ q_{out}^{f_1} = q_{in}^{f_2}, \ \ldots, q_{out}^{f_{n-1}} = q_{in}^{f_n}, \ q_{out}^{f_n} = q_{in}^{e'} \right\} \,.$$

By definition of Δ_T and $\Delta_{T/F}$, the projection $M^{1+2k(T)+d} \to M^{1+2k(T/F)+d}$ which projects away from the components associated with elements of F maps this space diffeomorphically onto

$$\Delta_{T/F} = \left\{ (q_0, (q_{in}^e, q_{out}^e)_{e \in E_{int}(T/F)}, q_1, \ldots, q_d) \in M^{1+2k(T/F)+d} \right.$$
$$\left| \ q_{out}^e = q_{in}^f \text{ for every } e, f \in E_{int}(T/F) \text{ with } v_{out}(e) = v_{in}(f) \,, \right.$$
$$q_{out}^e = q_j \text{ for every } e \in E_{int}(T/F), \ j \in \{1, \ldots, d\} \text{ with } v_{out}(e) = v_{in}(e_j(T/F)) \,,$$
$$\left. q_0 = q_{in}^e \text{ for every } e \in E_{int}(T/F) \text{ with } v_{out}(e_0(T/F)) = v_{in}(e) \right\} \,.$$

Thus, condition (5.4) implies

$$\left(\gamma_{0\infty}(0), (\gamma_{e\infty}(0), \gamma_{e\infty}(l_{e\infty}))_{e \in E_{int}(T)\setminus F}, \gamma_{1\infty}(0), \ldots, \gamma_{d\infty}(0)\right)$$
$$= \lim_{n\to\infty} \left(\gamma_{0n}(0), (\gamma_{en}(0), \gamma_{en}(l_{en}))_{e \in E_{int}(T)\setminus F}, \gamma_{1n}(0), \ldots, \gamma_{dn}(0)\right) \in \Delta_{T/F}. \quad (5.5)$$

Furthermore, since $\lim_{n\to\infty} l_{fn} = 0$ for every $f \in F$, we obtain

$$\gamma_{0\infty} \in W^- \left(x_0, (Y_0/F)\left((l_e)_{e \in E_{int}(T/F)}\right)\right),$$
$$(l_{e\infty}, \gamma_{e\infty}) \in \mathcal{M}\left((Y_e/F)\left((l_f)_{f \in E_{int}(T/F)\setminus\{e\}}\right)\right) \; \forall e \in E_{int}(T/F),$$
$$\gamma_{i\infty} \in W^+ \left(x_i, (Y_i/F)\left((l_e)_{e \in E_{int}(T/F)}\right)\right).$$

Together with (5.5), these observations yield that

$$\left(\gamma_{0\infty}, (l_{e\infty}, \gamma_{e\infty})_{e \in E_{int}(T)\setminus F}, \gamma_{1\infty}, \ldots, \gamma_{d\infty}\right) \in \mathcal{A}^d_{\pi_F(Y_T)}(x_0, x_1, \ldots, x_d, T/F).$$

Finally, we make use of the universality property. Up to this point, we only know that the limit is a perturbed Morse ribbon tree modelled on T/F, but with respect to a perturbation which is induced by Y_T.

For a general perturbation datum, this perturbation is not related to $Y_{T/F}$. But since the d-perturbation datum Y is universal, it holds that $\pi_F(Y_T) = Y_{T/F}$, which shows the claim.　　　　□

5.5　Geometric Convergence Along External Edges

The following theorem describes the case that all sequences of curves associated with internal edges of the tree converge and that all but one of the sequences associated with external edges converge while the remaining one is geometrically convergent. Formally speaking, it gives a precise description of the cases

$$E_1 = E(T) \setminus \{e_i(T)\}, \quad E_2 = \{e_i(T)\}, \quad E_3 = \emptyset, \quad \text{for some } i \in \{0, 1, \ldots, d\},$$

in Theorem 5.10.

Theorem 5.16 *Let $d \geq 2$, $T \in \mathrm{RTree}_d$ and $Y \in \mathfrak{X}(T)$. Let $x_0, x_1, \ldots, x_d \in \mathrm{Crit}\, f$ and let*

$$\left\{\left(\gamma_{0n}, (l_{en}, \gamma_{en})_{e \in E_{int}(T)}, \gamma_{1n}, \ldots, \gamma_{dn}\right)\right\}_{n \in \mathbb{N}} \subset \mathcal{A}^d_Y(x_0, x_1, \ldots, x_d, T)$$

such that

$$\liminf_{n\to\infty} l_{en} > 0 \; \forall e \in E_{int}(T).$$

For $e \in E(T)$ and $n \in \mathbb{N}$ we put

$$\bar{\gamma}_{en} := \begin{cases} (l_{en}, \gamma_{en}) & \text{if } e \in E_{int}(T), \\ \gamma_{in} & \text{if } e = e_i(T), \ i \in \{0, 1, \ldots, d\}. \end{cases}$$

For any such e and i we further put $(l_{e\infty}, \gamma_{e\infty}) := \lim_{n\to\infty}(l_{en}, \gamma_{en})$ *if* $\{(l_{en}, \gamma_{en})\}_{n\in\mathbb{N}}$ *converges and* $\gamma_{i\infty} := \lim_{n\to\infty} \gamma_{in}$ *if* $\{\gamma_{in}\}_{n\in\mathbb{N}}$ *converges.*

1. *Assume that the sequence* $\{\bar{\gamma}_{en}\}_{n\in\mathbb{N}}$ *converges for every* $e \in E(T) \setminus \{e_0(T)\}$ *and that* $\{\gamma_{0n}\}_{n\in\mathbb{N}}$ *converges geometrically against some*

$$(\hat{g}_1, \ldots, \hat{g}_{m-1}, \gamma_-) \in \widehat{\mathcal{M}}(x_0, y_1) \times \prod_{j=1}^{m-1} \widehat{\mathcal{M}}(y_j, y_{j+1}) \times W^-\left(y_m, Y_0\left(\vec{l}_\infty\right)\right),$$

where $\vec{l}_\infty = (l_{e\infty})_{e\in E_{int}(T)} \in (0, +\infty)^{E_{int}(T)}$. *Then*

$$\left(\gamma_-, (l_{e\infty}, \gamma_{e\infty})_{e\in E_{int}(T)}, \gamma_{1\infty}, \ldots, \gamma_{d\infty}\right) \in \mathcal{A}_{\mathbf{Y}}^d(y_m, x_1, \ldots, x_d, T).$$

2. *Assume that there is an* $i \in \{1, \ldots, d\}$ *such that the sequence* $\{\bar{\gamma}_{en}\}_{n\in\mathbb{N}}$ *converges for every* $e \in E(T) \setminus \{e_i(T)\}$ *and that* $\{\gamma_{in}\}_{n\in\mathbb{N}}$ *converges geometrically against some*

$$(\gamma_+, \hat{g}_1, \ldots, \hat{g}_m) \in W^+\left(y_1, Y_i\left(\vec{l}_\infty\right)\right) \times \prod_{j=1}^{m-1} \widehat{\mathcal{M}}(y_j, y_{j+1}) \times \widehat{\mathcal{M}}(y_m, x_i),$$

where $\vec{l}_\infty = (l_{e\infty})_{e\in E_{int}(T)} \in (0, +\infty)^{E_{int}(T)}$. *Then*

$$\left(\gamma_{0\infty}, (l_{e\infty}, \gamma_{e\infty})_{e\in E_{int}(T)}, \gamma_{1\infty}, \ldots, \gamma_{(i-1)\infty}, \gamma_+, \gamma_{(i+1)\infty}, \ldots, \gamma_{d\infty}\right)$$
$$\in \mathcal{A}_{\mathbf{Y}}^d(x_0, x_1, \ldots, x_{i-1}, y_1, x_{i+1}, \ldots, x_d, T).$$

See Fig. 5.4 for an illustration of the geometric limits in both parts of Theorem 5.16. The left-hand picture illustrates the first part, while the right-hand picture illustrates the second part.

Proof We first note that in both parts of the theorem, it obviously holds that

$$\lim_{n\to\infty} Y_i\left((l_{en})_{e\in E_{int}(T)}\right) = Y_i\left(\vec{l}_\infty\right) \quad \text{in } \mathfrak{X}^{lo}(T) \ \forall i \in \{0, 1, \ldots, d\}. \tag{5.6}$$

1. By (5.6) and part 1 of Theorem 5.1, $\{\gamma_{0n}\}_{n\in\mathbb{N}}$ converges to γ_- in the C_{loc}^∞-topology, which especially implies:
$$\lim_{n\to\infty} \gamma_{0n}(0) = \gamma_-(0).$$

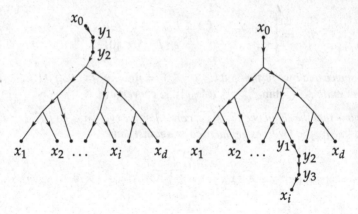

Fig. 5.4 The geometric convergence in Theorem 5.16

This has the following consequence:

$$\left(\gamma_-(0), (\gamma_{e\infty}(0), \gamma_{e\infty}(l_{e\infty}))_{e\in E_{int}(T)}, \gamma_{1\infty}(0), \ldots, \gamma_{d\infty}(0)\right)$$
$$= \lim_{n\to\infty} \left(\gamma_{0n}(0), (\gamma_{en}(0), \gamma_{en}(l_{en}))_{e\in E_{int}(T)}, \gamma_{1n}(0), \ldots, \gamma_{dn}(0)\right) \in \Delta_T,$$

which yields that

$$\left(\gamma_-, (l_{e\infty}, \gamma_{e\infty})_{e\in E_{int}(T)}, \gamma_{1\infty}, \ldots, \gamma_{d\infty}\right) \in \mathcal{A}_{\mathbf{Y}}^d(y_m, x_1, \ldots, x_d, T).$$

2. By (5.6) and part 2 of Theorem 5.1, $\{\gamma_{in}\}_{n\in\mathbb{N}}$ converges to γ_+ in the C_{loc}^∞-topology, which especially implies

$$\lim_{n\to\infty} \gamma_{in}(0) = \gamma_+(0).$$

This has the following consequence:

$$\left(\gamma_{0\infty}(0), (\gamma_{e\infty}(0), \gamma_{e\infty}(l_{e\infty}))_e, \gamma_{1\infty}(0), \ldots, \gamma_{(i-1)\infty}(0), \gamma_+(0), \gamma_{(i+1)\infty}(0), \ldots, \gamma_{d\infty}(0)\right)$$
$$= \lim_{n\to\infty} \left(\gamma_{0n}(0), (\gamma_{en}(0), \gamma_{en}(l_{en}))_{e\in E_{int}(T)}, \gamma_{1n}(0), \ldots, \gamma_{dn}(0)\right) \in \Delta_T,$$

which yields

$$\left(\gamma_{0\infty}, (l_{e\infty}, \gamma_{e\infty})_{e\in E_{int}(T)}, \gamma_{1\infty}, \ldots, \gamma_{(i-1)\infty}, \gamma_+, \gamma_{(i+1)\infty}, \ldots, \gamma_{d\infty}\right)$$
$$\in \mathcal{A}_{\mathbf{Y}}^d(x_0, x_1, \ldots, x_{i-1}, y_1, x_{i+1}, \ldots, x_d, T).$$

$\square$

5.6   Geometric Convergence Along Internal Edges

The next theorem covers the last of the special cases of convergence that we are considering. It discusses the case of geometric convergence of a family of sequences associated to internal edges of a ribbon tree while all other sequences are converging. This corresponds to the case

$$E_1 = E(T) \setminus E_2, \quad E_2 \subset E_{int}(T), \quad E_3 = \emptyset,$$

in Theorem 5.10.

Remember that we have equipped $E_{int}(T)$ with a fixed ordering for every $T \in$ RTree$_d$, which is necessary to make sense of the identification

$$\mathfrak{X}(T) = \mathfrak{X}(1, k(T), d) .$$

So for any $F \subset E_{int}(T)$ we can view F as a subset of $\{1, 2, \ldots, k\}$, with $k := k(T)$. In view of this identification we consider for $Y \in \mathfrak{X}_*(M, k)$ and $F = \{i_1, \ldots, i_d\}$ the vector field

$$Y_F \in \mathfrak{X}_\pm(M, k - |F|) ,$$
$$Y_F := \lim_{\lambda_1 \to +\infty} \lim_{\lambda_2 \to +\infty} \ldots \lim_{\lambda_{i_d} \to +\infty} c_{i_1} \left(\lambda_1, c_{i_2} \left(\lambda_2, \ldots, c_{i_d} (\lambda_d, Y) \ldots \right) \right) , \quad (5.7)$$

where $\mathfrak{X}_*(M, k)$ denotes one of the three spaces $\mathfrak{X}_-(M, k)$, $\mathfrak{X}_+(M, k)$ and $\mathfrak{X}_0(M, k)$. Here, $c_i(\lambda, Z)$ again denotes the contraction map given by inserting λ into the i-th parameter component of Z.

Since Y is of class C^{n+1} and the limits in (5.7) exist by definition of $\mathfrak{X}_\pm(M, k)$, the parametrized vector field Y_F is well defined and actually independent of the order of the limits in (5.7).

Theorem 5.17 *Let $d \geq 2$, $T \in$ RTree$_d$ and $\mathbf{Y} = (Y_0, (Y_e)_{e \in E_{int}(T)}, Y_1, \ldots, Y_d) \in \mathfrak{X}(T)$. Let $x_0, x_1, \ldots, x_d \in$ Crit f and*

$$\left\{ \left(\gamma_{0n}, (l_{en}, \gamma_{en})_{e \in E_{int}(T)} , \gamma_{1n}, \ldots, \gamma_{dn} \right) \right\}_{n \in \mathbb{N}} \subset \mathcal{A}_{\mathbf{Y}}^d (x_0, x_1, \ldots, x_d, T)$$

be a sequence with

$$\liminf_{n \to \infty} l_{en} > 0 \quad \forall e \in E_{int}(T) .$$

Assume that there is $F \subset E_{int}(T)$, such that for every $f \in F$ the sequence $\left\{ (l_{fn}, \gamma_{fn}) \right\}_{n \in \mathbb{N}}$ converges geometrically against some

$$\left(\gamma_{f+}, \hat{g}_1, \ldots, \hat{g}_m, \gamma_{f-}\right) \in W^+\left(y_{f+}, (Y_{f+})_F\left(\vec{l}_\infty\right)\right) \times \widehat{\mathcal{M}}(y_{f+}, y_{f1}) \times \prod_{j=1}^{m-2} \widehat{\mathcal{M}}(y_{fj}, y_{f(j+1)})$$

$$\times \widehat{\mathcal{M}}(y_{f(m-1)}, y_{f-}) \times W^-\left(y_{f-}, (Y_{f-})_F\left(\vec{l}_\infty\right)\right),$$

where

$$(Y_{f+}, Y_{f-}) = \lim_{\lambda \to \infty} \text{split}_{k(T)-1}(\lambda, Y_f),$$

and that $\{\bar{\gamma}_{en}\}_{n\in\mathbb{N}}$ *converges for every* $e \in E(T) \setminus F$, *where* $\bar{\gamma}_{en}$ *is defined as in Theorem 5.16. Then, using the notation from Theorem 3.10):*

$$\left(\gamma_{0\infty}, (\gamma_{f-})_{f\in F}, (l_{e\infty}, \gamma_{e\infty})_{e\in E_{int}(T)\setminus F}, \gamma_{1\infty}, \ldots, \gamma_{d\infty}, (\gamma_{f+})_{f\in F}\right)$$

$$\in \mathcal{M}_{\mathbf{Y}_F}\left(\left(x_0, (y_{f-})_{f\in F}\right), \left(x_1, \ldots, x_d, (y_{f+})_{f\in F}\right), (1+|F|, k(T)-|F|, d+|F|),\right.$$

$$\left.(\mathbb{R}_{>0})^{k(T)-|F|}, \sigma_F(\Delta_T)\right),$$

for some diffeomorphism

$$\sigma_F : M^{1+2k(T)+d} \stackrel{\cong}{\to} M^{1+2k(T)+d},$$

which is a permutation of the different factors of the product manifold, and where we put

$$\mathbf{Y}_F \in \mathfrak{X}(1+|F|, k(T)-|F|, d+|F|),\tag{5.8}$$

$$\mathbf{Y}_F := \left((Y_0)_F, ((Y_{f-})_F)_{f\in F}, ((Y_e)_F)_{e\in E_{int}(T)\setminus F}, (Y_1)_F, \ldots, (Y_d)_F, ((Y_{f+})_F)_{f\in F}\right).$$

See Fig. 5.5 for an illustration of the geometric convergence in Theorem 5.17.

Proof Let $f \in F$. By Theorem 5.6, we know that $\left\{\gamma_{fn}|_{[0,T]}\right\}_{n \geq n_T}$ converges against $\gamma_{f+}|_{[0,T]}$ in the C^∞-topology for every $T \geq 0$ and sufficiently big $n_T \in \mathbb{N}$, such that the restrictions are well-defined. This especially implies that

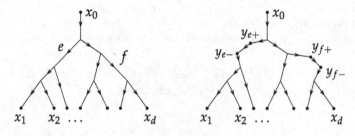

Fig. 5.5 Geometric convergence along $F = \{e, f\}$

$$\lim_{n\to\infty} \gamma_{fn}(0) = \gamma_{f+}(0) \,. \tag{5.9}$$

Moreover, we know by Theorem 5.6 that $\left\{\gamma_{fn}(\cdot + l_{fn})|_{[-T,0]}\right\}_{n\in\mathbb{N}}$ converges against $\gamma_{f-}|_{[-T,0]}$ in the C^∞-topology for every $T \geq 0$ and sufficiently big $n_T \in \mathbb{N}$, such that the restrictions are well-defined, which yields

$$\lim_{n\to\infty} \gamma_{fn}(l_n) = \gamma_{f-}(0) \,. \tag{5.10}$$

By definition of $\mathcal{A}_Y^d(x_0, x_1, \ldots, x_d, T)$, the following holds for every $n \in \mathbb{N}$:

$$\left(\gamma_{0n}(0), (\gamma_{en}(0), \gamma_{en}(l_{en}))_{e\in E_{int}(T)}, \gamma_{1n}(0), \ldots, \gamma_{dn}(0)\right) \in \Delta_T \,.$$

Thus, Eqs. (5.9) and (5.10) imply

$$\left(\gamma_{0\infty}(0), (\gamma_{f-}(0))_{f\in F}, (\gamma_{en}(0), \gamma_{en}(l_{en}))_{e\in E_{int}(T)\backslash F}, \gamma_{1\infty}(0), \ldots, \gamma_{d\infty}(0), (\gamma_{f+}(0))_{f\in F}\right)$$

$$\in \left\{ \left(q_0, \left(q_{in}^f\right)_{f\in F}, (q_{in}^e, q_{out}^e)_{e\in E_{int}(T)\backslash F}, q_1, \ldots, q_d, \left(q_{out}^f\right)_{f\in F}\right) \in M^{1+2k(T)+d} \right.$$

$$\left. \left| \left(q_0, (q_{in}^e, q_{out}^e)_{e\in E_{int}(T)}, q_1, \ldots, q_d\right) \in \Delta_T \right\} \,.$$

Obviously, there is a permutation of the factors $\sigma_F : M^{1+2k(T)+d} \to M^{1+2k(T)+d}$ which maps Δ_T diffeomorphically onto this set. The claim immediately follows. $\square$

Definition 5.18 In the situation of Theorem 5.17, we define:

$$\mathcal{B}_Y\left((x_0, (y_{e-})_{e\in F}), (x_1, \ldots, x_d, (y_{e+})_{e\in F}), T, F\right)$$
$$:= \mathcal{M}_{Y_F}\left((x_0, (y_{e-})_{e\in F}), (x_1, \ldots, x_d, (y_{e+})_{e\in F}), (1+|F|, k(T)-|F|, d+|F|), \right.$$
$$\left. (0, +\infty)^{k(T)-|F|}, \sigma_F(\Delta_T)\right) \,.$$

Before we conclude this section by a general theorem describing limit spaces of sequences of perturbed Morse ribbon trees and a final regularity proposition, we consider the following regularity result for the spaces from Definition 5.18.

Proposition 5.19 Let $d \geq 2$, $T \in \text{RTree}_d$. For generic choice of $Y \in \mathfrak{X}(T)$, the following holds: For all $F \subset E_{int}(T)$ and $x_0, x_1, \ldots, x_d, y_0, y_1, \ldots, y_d \in \text{Crit } f$ with

$$\mu(x_0) \geq \mu(y_0) \quad \text{and} \quad \mu(y_i) \geq \mu(x_i) \quad \text{for every } i \in \{1, 2, \ldots, d\} \,, \tag{5.11}$$

and for all $e \in F$ and $y_{e+}, y_{e-} \in \operatorname{Crit} f$ with

$$\mu(y_{e+}) \geq \mu(y_{e-}), \tag{5.12}$$

the spaces

$$\mathcal{A}_{\mathbf{Y}}^d(x_0, x_1, \ldots, x_d, T) \quad and \quad \mathcal{B}_{\mathbf{Y}}\left((y_0, (y_{e-})_{e \in F}), (y_1, \ldots, y_d, (y_{e+})_{e \in F}), T, F\right)$$

are manifolds of class C^{n+1}. The following inequality is true for any of these choices:

$$\dim \mathcal{B}_{\mathbf{Y}}\left((y_0, (y_{e-})_{e \in F}), (y_1, \ldots, y_d, (y_{e+})_{e \in F}), T, F\right) \leq \dim \mathcal{A}_{\mathbf{Y}}^d(x_0, x_1, \ldots, x_d, T) - |F|.$$

Proof Let $F \subset E_{int}(T)$. It follows immediately from Theorem 3.10 that there is a generic subset

$$\mathcal{G}_F \subset \mathfrak{X}(1 + |F|, k(T) - |F|, d + |F|)$$

such that for every $\mathbf{Z} \in \mathcal{G}_F$, the space

$$\mathcal{M}_{\mathbf{Z}}\left((y_0, (y_{e-})_{e \in F}), (y_1, \ldots, y_d, (y_{e+})_{e \in F}), (1 + |F|, k(T) - |F|, d + |F|),\right.$$
$$\left. (0, +\infty)^{k(T) - |F|}, \sigma_F(\Delta_T)\right)$$

is a manifold of class C^{n+1}. Furthermore, Theorem 3.10 implies the existence of a generic subset $\mathcal{G} \subset \mathfrak{X}(T)$ such that for every $\mathbf{Y} \in \mathcal{G}$, the space $\mathcal{A}_{\mathbf{Y}}^d(x_0, \ldots, x_d, T)$ is a manifold of class C^{n+1}.

To compare these perturbations, note that the map:

$$p_F : \mathfrak{X}(T) \to \mathfrak{X}(1 + |F|, k(T) - |F|, d + |F|), \quad \mathbf{Y} \mapsto \mathbf{Y}_F,$$

where $\mathbf{Y}_F$ is defined as in (5.8), is continuous and surjective. Therefore, the set $p_F^{-1}(\mathcal{G}_F)$ is generic in $\mathfrak{X}(T)$. It follows that the regularity statement holds for every

$$\mathbf{Y} \in \bigcap_{F \subset E_{int}(T)} p_F^{-1}(\mathcal{G}_F) \cap \mathcal{G},$$

and since $\bigcap_{F \subset E_{int}(T)} p_F^{-1}(\mathcal{G}_F) \cap \mathcal{G}$ is a finite intersection of generic sets, it is itself generic in $\mathfrak{X}(T)$.

It remains to show the dimension inequality. By Theorem 3.10, the dimension is computed as follows:

$$\dim \mathcal{B}_{\mathbf{Y}}\left((y_0, (y_{e-})_{e \in F}), (y_1, \ldots, y_d, (y_{e+})_{e \in F}), T, F\right)$$

$$= \dim \mathcal{M}_{\mathbf{Y}_F}\left((y_0, (y_{e-})_{e \in F}), (y_1, \ldots, y_d, (y_{e+})_{e \in F}), (1 + |F|, k(T) - |F|, d + |F|),\right.$$

$$\left. (0, +\infty)^{k(T)-|F|}, \sigma_F(\Delta_T)\right)$$

$$= \mu(y_0) + \sum_{e \in F} \mu(y_{e-}) - \sum_{i=1}^d \mu(y_i) - \sum_{e \in F} \mu(y_{e+}) + (k(T) - |F| + d + |F|)n + k(T) - |F|$$

$$- \operatorname{codim} \sigma_F(\Delta_T)$$

$$= \mu(y_0) - \sum_{i=1}^d \mu(y_i) + \sum_{e \in F} (\mu(y_{e-}) - \mu(y_{e+})) + (k(T) + d)n + k(T) - |F| - \operatorname{codim} \Delta_T \ .$$

Applying formula (4.3) for the codimension of Δ_T, we obtain:

$$\dim \mathcal{B}_{\mathbf{Y}}\left((y_0, (y_{e-})_{e \in F}), (y_1, \ldots, y_d, (y_{e+})_{e \in F}), T, F\right)$$

$$= \mu(y_0) - \sum_{i=1}^d \mu(y_i) + \sum_{e \in F} (\mu(y_{e-}) - \mu(y_{e+})) + k(T) - |F| \ .$$

Using assumptions (5.11) and (5.12), we derive:

$$\dim \mathcal{B}_{\mathbf{Y}}\left((y_0, (y_{e-})_{e \in F}), (y_1, \ldots, y_d, (y_{e+})_{e \in F}), T, F\right) \le \mu(x_0) - \sum_{i=1}^d \mu(x_i) + k(T) - |F|$$

$$= \dim \mathcal{A}_{\mathbf{Y}}(x_0, x_1, \ldots, x_d) - |F| \ .$$

$\square$

Note that by Theorem 5.10, the different convergence phenomena described in Theorems 5.15, 5.16 and 5.17 can occur simultaneously.

5.7 Sequential Compactness and the Regularity of Boundary Spaces

The upcoming Theorem 5.22 subsumizes all convergence phenomena for sequences of perturbed Morse ribbon trees by stating a general compactness property of the moduli spaces $\mathcal{A}_{\mathbf{Y}}^d(x_0, x_1, \ldots, x_d, T)$. It completes our discussion of geometric convergence phenomena for these sequences and is shown by applying the arguments used to prove the aforementioned theorems simultaneously. We omit the details.

Definition 5.20 Let $T \in \operatorname{RTree}_d$, $d \ge 2$, $Y \in \mathfrak{X}(T)$, $x_0, x_1, \ldots, x_d \in \operatorname{Crit} f$ and let

$$\left\{\left(\gamma_{0n}, (l_{en}, \gamma_{en})_{e \in E_{int}(T)}, \gamma_{1n}, \ldots, \gamma_{dn}\right)\right\}_{n \in \mathbb{N}} \subset \mathcal{A}_{\mathbf{Y}}^d(x_0, x_1, \ldots, x_d, T)$$

be a sequence of perturbed Morse ribbon trees.

The sequence *converges in* $\mathcal{A}_{\mathbf{Y}}^d(x_0, x_1, \ldots, x_d, T)$ if all of its component sequences $\{\gamma_{in}\}_n$, $i \in \{0, 1, \ldots, d\}$, and $\{(l_{en}, \gamma_{en})\}_n$, $e \in E_{int}(T)$, converge and if the product of the limits of the component sequences lies in $\mathcal{A}_{\mathbf{Y}}^d(x_0, x_1, \ldots, x_d, T)$.

The sequence *converges geometrically* if every component sequence is either convergent or geometrically convergent and at least one of the following holds true:

- there exists a component sequence that converges geometrically,
- there exists an $e \in E_{int}(T)$ with $\lim_{n \to \infty} l_{en} = 0$.

Remark 5.21 Since by definition of $\mathcal{A}_{\mathbf{Y}}^d(x_0, x_1, \ldots, x_d, T)$, the components of its elements include only finite-length trajectories (l, γ) with $l > 0$, it follows that a sequence

$$\left\{\left(\gamma_{0n}, (l_{en}, \gamma_{en})_{e \in E_{int}(T)}, \gamma_{1n}, \ldots, \gamma_{dn}\right)\right\}_{n \in \mathbb{N}} \subset \mathcal{A}_{\mathbf{Y}}^d(x_0, x_1, \ldots, x_d, T),$$

whose component sequences are all convergent, converges in $\mathcal{A}_{\mathbf{Y}}^d(x_0, x_1, \ldots, x_d, T)$ if and only if

$$\lim_{n \to \infty} l_{en} > 0 \quad \forall e \in E_{int}(T).$$

Theorem 5.22 *Let* $d \geq 2$, $T \in \mathrm{RTree}_d$, $\mathbf{Y} \in \mathfrak{X}(T)$ *and* $x_0, x_1, \ldots, x_d \in \mathrm{Crit}\, f$. *Let*

$$\left\{\underline{\gamma}_n\right\}_{n \in \mathbb{N}} = \left\{\left(\gamma_{0n}, (l_{en}, \gamma_{en})_{e \in E_{int}(T)}, \gamma_{1n}, \ldots, \gamma_{dn}\right)\right\}_{n \in \mathbb{N}}$$

be a sequence in $\mathcal{A}_{\mathbf{Y}}^d(x_0, x_1, \ldots, x_d, T)$ *which does not have a convergent subsequence. Then there are sets*

$$F_1, F_2 \subset E_{int}(T), \quad F_1 \cap F_2 = \emptyset$$

and a subsequence $\left\{\underline{\gamma}_{n_k}\right\}_{k \in \mathbb{N}}$, *such that the following holds:*

Up to a permutation of its components, $\left\{\underline{\gamma}_{n_k}\right\}_{k \in \mathbb{N}}$ *converges geometrically against a product of unparametrized Morse trajectories and an element of*

$$\mathcal{B}_{\mathbf{Y}}\left(\left(y_0, (y_{e-})_{e \in F_1}\right), \left(y_1, \ldots, y_d, (y_{e+})_{e \in F_1}\right), T/F_2, F_1\right),$$

where $y_0, y_1, \ldots, y_d \in \mathrm{Crit}\, f$, $y_{e-}, y_{e+} \in \mathrm{Crit}\, f$ *for* $e \in F_1$ *satisfy*

$$\mu(y_0) \leq \mu(x_0), \quad \mu(y_i) \geq \mu(x_i)\, \forall i \in \{1, 2, \ldots, d\}, \quad \mu(y_{e+}) \geq \mu(y_{e-})\, \forall e \in F_1.$$
$$(5.13)$$

Moreover, for every inequality in (5.13) *equality holds if and only if the respective critical points are identical.*

We conclude this section by providing a regularity result for the moduli spaces occurring in Theorem 5.22.

Proposition 5.23 *For generic choice of* $\mathbf{Y} \in \mathfrak{X}(T)$*, it holds for all* $F_1, F_2 \subset E_{int}(T)$ *with* $F_1 \cap F_2 = \emptyset$*,* $y_0, y_1, \ldots, y_d \in \text{Crit } f$ *and* $y_{e+}, y_{e-} \in \text{Crit } f$ *for* $e \in F_1$ *satisfying (5.13) that the space*

$$\mathcal{B}_{\mathbf{Y}}\left(\left(y_0, (y_{e-})_{e \in F_1}\right), \left(y_1, \ldots, y_d, (y_{e+})_{e \in F_1}\right), T/F_2, F_1\right) \tag{5.14}$$

is a manifold of class C^{n+1}*. Moreover, the following inequality holds for any of these choices:*

$$\begin{aligned}
\dim \mathcal{B}_{\mathbf{Y}}&\left(\left(y_0, (y_{e-})_{e \in F_1}\right), \left(y_1, \ldots, y_d, (y_{e+})_{e \in F_1}\right), T/F_2, F_1\right) \\
&\leq \dim \mathcal{A}_{\mathbf{Y}}^d(x_0, x_1, \ldots, x_d, T) - |F_1| - |F_2|.
\end{aligned} \tag{5.15}$$

Proof The statement follows almost immediately from Proposition 5.19.

By Proposition 5.19, we can find a generic subset $\mathcal{G}_{F_2}$ for any $F_2 \subset E_{int}(T)$, such that the space in (5.14) is a manifold of class C^{n+1} for any $F_1 \subset E(T/F_2) = E(T) \setminus F_2$ and any choice of critical points.

Consider the maps $\pi_{F_2} : \mathfrak{X}(T) \to \mathfrak{X}(T/F_2)$ from (5.3). Since these maps are surjective and continuous, the set $\pi_{F_2}^{-1}(\mathcal{G}_{F_2})$ is a generic subset of $\mathfrak{X}(T)$ for every F_2. Therefore, the set

$$\bigcap_{F_2 \subset E_{int}(T)} \pi_{F_2}^{-1}(\mathcal{G}_{F_2})$$

is a generic subset of $\mathfrak{X}(T)$ having the desired properties.

Considering the inequality (5.15), note that by Proposition 5.19 the following holds for every $F_1, F_2 \subset E_{int}(T)$ as in the statement:

$$\begin{aligned}
\dim \mathcal{B}_{\mathbf{Y}}&\left(\left(y_0, (y_{e-})_{e \in F_1}\right), \left(y_1, \ldots, y_d, (y_{e+})_{e \in F_1}\right), T/F_2, F_1\right) \\
&\leq \dim \mathcal{A}_{\mathbf{Y}}^d(x_0, x_1, \ldots, x_d, T/F_2) - |F_1| \\
&= \mu(x_0) - \sum_{i=1}^{d} \mu(x_i) + k(T) - |F_1| - |F_2| \\
&= \dim \mathcal{A}_{\mathbf{Y}}^d(x_0, x_1, \ldots, x_d, T) - |F_1| - |F_2|,
\end{aligned}$$

where we use the dimension formula from Theorem 4.15 and the fact that

$$k(T/F_2) = k(T) - |F_2|.$$

$\square$

Definition 5.24 Let $d \geq 2$. A d-perturbation datum $\mathbf{Y} = (\mathbf{Y}_T)_{T \in \text{RTree}_d}$ is called *regular* if for every $T \in \text{RTree}_d$ and every $F_1, F_2 \subset E_{int}(T)$ with $F_1 \cap F_2 = \emptyset$ the spaces

$$\mathcal{B}_{\mathbf{Y}}\left(\left(y_0, (y_{e-})_{e \in F_1}\right), \left(y_1, \dots, y_d, (y_{e+})_{e \in F_1}\right), T/F_2, F_1\right) \quad \text{and} \quad \mathcal{A}_{\mathbf{Y}}^d(x_0, x_1, \dots, x_d, T)$$

are manifolds of class C^{n+1} for all $x_0, \dots, x_d, y_0, \dots, y_d \in \mathrm{Crit}\, f$ and $y_{e-}, y_{e+} \in \mathrm{Crit}\, f, e \in F_1$.

Proposition 5.23 and Theorem 4.15 together imply that for any $d \geq 2$, the set of regular d-perturbation data is generic in the space of all d-perturbation data $\prod_{T \in \mathrm{RTree}_d} \mathfrak{X}(T)$.

Chapter 6
Higher Order Multiplications and the A_∞-Relations

We continue by taking a closer look at zero- and one-dimensional moduli spaces of perturbed Morse ribbon trees. The results from Chap. 5 enable us to show that zero-dimensional moduli spaces are in fact finite sets. This basic observation allows us to define homomorphisms $C^*(f)^{\otimes d} \to C^*(f)$ for every $d \geq 2$ via counting elements of these zero-dimensional moduli spaces. After constructing these higher order multiplications explicitly, we will study the compactifications of one-dimensional moduli spaces of perturbed Morse ribbon trees. The results from Chap. 5 imply that one-dimensional moduli spaces can be compactified to one-dimensional manifolds with boundary, and we will explicitly describe their boundaries. If we impose an additional consistency condition on the chosen perturbations, the boundary spaces will coincide with product of zero-dimensional moduli spaces of perturbed Morse ribbon trees.

We will be able to show via counting elements of these boundaries that the higher order multiplications indeed satisfy the defining equations of an A_∞-algebra. The consistency condition will be formulated in terms of the background perturbations that we introduced in Chaps. 3 and 4.

Throughout this chapter, we assume again that every ribbon tree is equipped with an ordering of its internal edges.

6.1 An Existence Result for Regular Perturbations

We have introduced the notions of regular and universal perturbation data in Chap. 5. It will turn out that the perturbation data which allow us to draw the desired consequences are those which are *both* regular *and* universal. It is easy to see that there are perturbation data with either of these properties, but it is not obvious that there are indeed perturbation data with both properties. Before focussing on zero- and one-dimensional moduli spaces of perturbed Morse ribbon trees, we therefore begin this chapter with a nontrivial existence result.

© Springer International Publishing AG, part of Springer Nature 2018
S. Mescher, *Perturbed Gradient Flow Trees and A∞-algebra Structures in Morse Cohomology*, Atlantis Studies in Dynamical Systems 6,
https://doi.org/10.1007/978-3-319-76584-6_6

Lemma 6.1 *Let $d \geq 2$ and let $\mathbf{X} = (\mathbf{X}_T)_{T \in \mathrm{RTree}_d^*}$ be a universal background d-perturbation datum. Then there exists a regular and universal d-perturbation datum $\mathbf{Y} = (\mathbf{Y}_T)_{T \in \mathrm{RTree}_d}$ with $\mathbf{Y}_T \in \mathfrak{X}(T, \mathbf{X}_T)$ for every $T \in \mathrm{RTree}_d$.*

Proof One computes that since $\mathbf{X}$ is universal, the map $\pi_F : \mathfrak{X}(T) \to \mathfrak{X}(T/F)$ restricts to a map

$$\pi_F|_{\mathfrak{X}(T, \mathbf{X}_T)} : \mathfrak{X}(T, \mathbf{X}_T) \to \mathfrak{X}(T/F, \mathbf{X}_{T/F}) \, ,$$

for all $T \in \mathrm{RTree}_d$ and $F \subset E_{int}(T)$ with $k(T/F) > 0$, which is a necessary condition for such a universal d-perturbation datum to exist. We will prove the claim by inductively constructing regular and universal perturbation data over the number of internal edges of the trees.

Fix $d \geq 2$, let $T_0 \in \mathrm{RTree}_d$ be the unique d-leafed ribbon tree with $k(T_0) = 0$ and put $\mathfrak{X}(T, \mathbf{X}_T) := \mathfrak{X}(T)$. (See picture b) in Fig. 4.1.) For any $x_0, x_1, \ldots, x_d \in \mathrm{Crit}\, f$, the space $\mathcal{A}_{\mathbf{Y}}^d(x_0, x_1, \ldots, x_d, T_0)$ is then a subset of a product of spaces of semi-infinite perturbed Morse trajectories only, so the conditions defining universality are irrelevant for perturbed Morse ribbon trees modelled on T_0. Moreover, the geometric limits of all sequences $\mathcal{A}_{\mathbf{Y}}^d(x_0, x_1, \ldots, x_d, T_0)$ lie in products of unparametrized trajectory spaces and a space of perturbed Morse ribbon trees which are again modelled on T_0. Therefore, if $\mathbf{Y}_{T_0} \in \mathfrak{X}(T_0)$ is regular, all possible boundary spaces of $\mathcal{A}_{\mathbf{Y}}^d(x_0, x_1, \ldots, x_d, T_0)$ are again smooth manifolds.

As induction hypothesis, we assume that we have found a regular perturbation $\mathbf{Y}_T \in \mathfrak{X}(T, \mathbf{X}_T)$ for every $T \in \mathrm{RTree}_d$ with $k(T) \leq k$ for some fixed $k \in \{0, 1, \ldots, d - 3\}$, such that

$$\pi_F(\mathbf{Y}_T) = \mathbf{Y}_{T/F}$$

for every $F \subset E_{int}(T)$ and such that every boundary space of the form

$$\mathcal{B}_{\mathbf{Y}} \left(\left(y_0, (y_{e-})_{e \in F_1} \right), \left(y_1, \ldots, y_d, (y_{e+})_{e \in F_1} \right), T/F_2, F_1 \right)$$

appearing in Theorem 5.22 is a manifold of class C^{n+1}. (These assumptions are well-defined, since $k(T) \leq k$ implies $k(T/F) \leq k$ for every $F \subset E_{int}(T)$.)

Loosely speaking, this means that we assume that we have chosen a family of perturbations which is regular and universal for trees with *up to k internal edges*.

Let now $T_1 \in \mathrm{RTree}_d$ with $k(T_1) = k + 1$ and consider the perturbation space

$$\tilde{\mathfrak{X}} := \left\{ \mathbf{Y}_1 \in \mathfrak{X}(T_1, \mathbf{X}_{T_1}) \mid \pi_F(\mathbf{Y}_1) = \mathbf{Y}_{T_1/F} \ \forall F \subset E_{int}(T_1) \right\} \, .$$

This is again well-defined since $k(T_1/F) \leq k$ for every non-empty $F \subset E_{int}(T_1)$, so $\mathbf{Y}_{T_1/F}$ has already been chosen. $\tilde{\mathfrak{X}}$ as a closed affine subspace of $\mathfrak{X}(T_1)$, whose underlying linear subspace is given by

$$\{ \mathbf{Y}_1 \in \mathfrak{X}(T_1, 0) \mid \pi_F(\mathbf{Y}_1) = 0 \ \forall F \subset E_{int}(T_1) \} \, ,$$

where 0 denotes the family consisting of vanishing vector fields in $\mathfrak{X}^{\text{back}}(T_1)$ and $\mathfrak{X}(T/F)$, respectively. Hence, $\widetilde{\mathfrak{X}}$ is a Banach submanifold of $\mathfrak{X}(T)$. Remember that the proof of the regularity statement in Theorem 4.15 is essentially an application of Theorem 3.10. By taking a closer look at the proof of Theorem 3.10 one checks without difficulty that the whole line of argument will still hold if we restrict to the (obviously non-empty) perturbation space $\widetilde{\mathfrak{X}}$. More precisely, there is a generic subset $\mathcal{G} \subset \widetilde{\mathfrak{X}}$, such that for $\mathbf{Y}_{T_1} \in \widetilde{\mathfrak{X}}$ the space $\mathcal{A}^d_{\mathbf{Y}_{T_1}}(x_0, x_1, \ldots, x_d, T_1)$ is a manifold of class C^{n+1} for all $x_0, x_1, \ldots, x_d \in \operatorname{Crit} f$.

Moreover, the same is true in proof of Proposition 5.19, i.e. we can again restrict to elements of $\widetilde{\mathfrak{X}}$ in the situation of this proposition. It follows that for generic choice of $\mathbf{Y}_{T_1} \in \widetilde{\mathfrak{X}}$, the space $\mathcal{A}^d_{\mathbf{Y}_{T_1}}(x_0, x_1, \ldots, x_d, T_1)$ as well as all boundary spaces of the form $\mathcal{B}_{\mathbf{Y}_{T_1}}(\ldots)$ are manifolds of class C^{n+1}.

For every $T_1 \in \text{RTree}_1$ we choose such a generic $\mathbf{Y}_{T_1} \in \mathfrak{X}(T_1, \mathbf{X}_{T_1})$. Then the family

$$(\mathbf{Y}_T)_{T \in \text{RTree}_d, \, k(T) \leq k+1}$$

satisfies the regularity and universality condition for every tree with up to $k+1$ internal edges. Proceeding inductively, we can therefore construct a regular and universal d-perturbation datum. $\square$

6.2   Zero-Dimensional Moduli Spaces of Perturbed Morse Ribbon Trees

To define higher order multiplications on the Morse cochain complexes on the Morse cochain complex of f, we need to study zero-dimensional moduli spaces of perturbed Morse ribbon trees. The following theorem will eventually allow us to define the multiplications.

Theorem 6.2 *Let* $\mathbf{Y} = (\mathbf{Y}_T)_{T \in \text{RTree}_d}$ *be a regular and universal d-perturbation datum, $d \geq 2$. Then for all $T \in \text{RTree}_d$ and $x_0, x_1, \ldots, x_d \in \operatorname{Crit} f$ with*

$$\mu(x_0) = \sum_{i=1}^{d} \mu(x_i) - k(T) \,, \tag{6.1}$$

the space $\mathcal{A}^d_{\mathbf{Y}}(x_0, x_1, \ldots, x_d, T)$ is a finite set.

Proof Since $\mathbf{Y}$ is a regular d-perturbation datum, Theorem 4.15 implies that the space $\mathcal{A}^d_{\mathbf{Y}}(x_0, x_1, \ldots, x_d, T)$ is a zero-dimensional manifold, hence a discrete set. To show that it is finite, it therefore suffices to show that $\mathcal{A}^d_{\mathbf{Y}}(x_0, x_1, \ldots, x_d, T)$ is sequentially compact, which we will do by using the results from Sect. 5.7. Assume that there is a sequence

$$\left\{ \left(\gamma_{0n}, (l_{en}, \gamma_{en})_{e \in E_{int}(T)}, \gamma_{1n}, \ldots, \gamma_{dn} \right) \right\}_{n \in \mathbb{N}} \subset \mathcal{A}^d_{\mathbf{Y}}(x_0, x_1, \ldots, x_d, T)$$

which does not have a subsequence converging in $\mathcal{A}_{\mathbf{Y}}^d(x_0, x_1, \ldots, x_d, T)$. By Theorem 5.22, there are only two cases that can occur:

1. The sequence has a subsequence for which all component sequences converge and there is $F \subset E_{int}(T)$, $F \neq \emptyset$, with $\lim_{n\to\infty} l_{en} = 0$ for every $e \in F$.

   By Theorem 5.15 and the universality of $\mathbf{Y}$, we can identify the limit with an element of $\mathcal{A}_{\mathbf{Y}}^d(x_0, x_1, \ldots, x_d, T/F)$. Since $\mathbf{Y}$ is a regular d-perturbation datum, $\mathcal{A}_{\mathbf{Y}}^d(x_0, x_1, \ldots, x_d, T/F)$ is a manifold of class C^{n+1} of expected dimension

$$\dim \mathcal{A}_{\mathbf{Y}}^d(x_0, x_1, \ldots, x_d, T/F) = \mu(x_0) - \sum_{i=1}^d \mu(x_i) + k(T/F)$$

$$= \mu(x_0) - \sum_{i=1}^d \mu(x_i) + k(T) - F \overset{(6.1)}{=} -F < 0,$$

   since F is non-empty. This yields

$$\mathcal{A}_{\mathbf{Y}}^d(x_0, x_1, \ldots, x_d, T/F) = \emptyset.$$

   Therefore, since $\mathbf{Y}$ is regular, the limit can not exist and the collapse of internal edges can not occur.

2. The sequence has a geometrically convergent subsequence, such that at least one of the component sequences converges geometrically.

   By Theorem 5.22, there exist $F_1, F_2 \subset E(T)$ with $F_1 \neq \emptyset$, such that the subsequence converges geometrically against an element of

$$\mathcal{B}_{\mathbf{Y}}\left((y_0, (y_{e-})_{e\in F_1}), (y_1, \ldots, y_d, (y_{e+})_{e\in F_1}), T/F_2, F_1\right)$$

   for some critical points having the same properties as in Theorem 5.22. But by Proposition 5.23, this space is a manifold of class C^{n+1} whose expected dimension fulfils:

$$\dim \mathcal{B}_{\mathbf{Y}}\left((y_0, (y_{e-})_{e\in F_1}), (y_1, \ldots, y_d, (y_{e+})_{e\in F_1}), T/F_2, F_1\right)$$
$$\leq \dim \mathcal{A}_{\mathbf{Y}}^d(x_0, x_1, \ldots, x_d, T/F_2) - |F_1|$$
$$= \mu(x_0) - \sum_{i=1}^d \mu(x_i) + k(T) - |F_1| - |F_2| \overset{(6.1)}{=} -|F_1| - |F_2| < 0,$$

   since F_1 is nonempty. Thus, the limit can not exist and this case of geometric convergence does not occur in our situation.

By Theorem 5.22, these two are the only possible cases for any such sequence. Therefore, every sequence in $\mathcal{A}_{\mathbf{Y}}^d(x_0, x_1, \ldots, x_d, T)$ has a convergent subsequence. We conclude that $\mathcal{A}_{\mathbf{Y}}^d(x_0, x_1, \ldots, x_d, T)$ is compact and discrete, hence finite, for any choice of $x_0, x_1, \ldots, x_d \in \mathrm{Crit}\, f$ satisfying (6.1). $\square$

Theorem 6.2 is of great importance, since it allows us to count elements of zero-dimensional moduli spaces and thereby construct maps, similar to the counting of elements of zero-dimensional spaces of unparametrized Morse trajectories to define the differentials of Morse (co)chain complexes, see [Sch93].

Before we do so, we need to consider the orientability of moduli spaces of perturbed Morse ribbon trees. The consideration of orientations on these moduli spaces will be necessary in order to construct higher order multiplications which satisfy the defining equations of an A_∞-algebra. Theorem 6.2 implies that an oriented zero-dimensional moduli space consists of a finite number of points which are each equipped with an orientation, i.e. a sign.

For this purpose, we want rephrase the question of regularity of moduli spaces of Morse ribbon trees explicitly as a transverse intersection problem. Constructing orientations on the spaces involved and using oriented intersection theory, we will therefore be able to define algebraic intersection numbers. Up to a minor modification, these algebraic intersection numbers will be the coefficients of the higher order multiplications.

For the necessary results from oriented intersection theory, we refer to the textbooks [Hir76, Sect. 5.2], [GP74, Sect. 3.3] and [BH04, Sect. 5.6].

Let $d \geq 2$, $T \in \mathrm{RTree}_d$ and $\mathbf{Y} = (Y_0, (Y_e)_{e \in E_{int}(T)}, Y_1, \ldots, Y_d) \in \mathfrak{X}(T)$. For given critical points $x_0, x_1, \ldots, x_d \in \mathrm{Crit}\, f$, consider the space

$$\mathcal{M}_{\mathbf{Y}}^d(x_0, x_1, \ldots, x_d, T) := \Big\{ (\gamma_0, (l_e, \gamma_e)_{e \in E_{int}(T)}, \gamma_1, \ldots, \gamma_d) \,\Big|\, \gamma_0 \in W^-(x_0, Y_0((l_e)_e)),$$
$$(l_e, \gamma_e) \in \mathcal{M}\big(Y_e\big((l_f)_{f \in E_{int}(T) \setminus \{e\}}\big)\big) \text{ and } l_e > 0 \quad \forall e \in E_{int}(T),$$
$$\gamma_i \in W^+\big(x_i, Y_i\big((l_e)_{e \in E_{int}(T)}\big)\big) \,\forall i \in \{1, 2, \ldots, d\} \Big\}.$$

Note that we introduced this space in the proof of Theorem 3.10 under the name $\overline{\mathcal{W}}_{\mathbf{Y}}$. As discussed in the proof of Theorem 3.10, $\mathcal{M}_{\mathbf{Y}}^d(x_0, x_1, \ldots, x_d, T)$ is a smooth manifold of dimension

$$\dim \mathcal{M}_{\mathbf{Y}}^d(x_0, x_1, \ldots, x_d, T) = \mu(x_0) - \sum_{i=1}^{d} \mu(x_i) + (k(T) + d)n + k(T).$$

Consider the map

$$\underline{E}_{\mathbf{Y}} : \mathcal{M}_{\mathbf{Y}}^d(x_0, x_1, \ldots, x_d, T) \to M^{1+2k(T)+d}, \tag{6.2}$$
$$(\gamma_0, (l_e, \gamma_e)_{e \in E_{int}(T)}, \gamma_1, \ldots, \gamma_d) \mapsto (\gamma_0(0), (\gamma_e(0), \gamma_e(l_e))_{e \in E_{int}(T)}, \gamma_1(0), \ldots, \gamma_d(0)),$$

which we also introduced in the proof of Theorem 3.10, see especially (3.8). By definition of regularity, the perturbation $\mathbf{Y}$ is regular if and only if the map $\underline{E}_{\mathbf{Y}}$ intersects the T-diagonal $\Delta_T \subset M^{1+2k(T)+d}$ transversely, and we can write

$$\mathcal{A}_{\mathbf{Y}}^d(x_0, x_1, \ldots, x_d, T) = \underline{E}_{\mathbf{Y}}^{-1}(\Delta_T).$$

The manifolds $\mathcal{M}_Y^d(x_0, x_1, \ldots, x_d, T)$ are orientable, but the definition of an orientation on $\mathcal{M}_Y^d(x_0, x_1, \ldots, x_d, T)$ requires further considerations of orientations on the three types of moduli spaces of (unperturbed) negative gradient flow trajectories we introduced in Chap. 2. We postpone these considerations to Appendix A.1.

The details about orienting the spaces $\mathcal{M}_Y^d(x_0, x_1, \ldots, x_d, T)$ are very technical and the interested reader can find the details in Appendix A.4.

The orientation of $\mathcal{M}_Y^d(x_0, x_1, \ldots, x_d, T)$ especially depends on the choice of ordering of $E_{int}(T)$.

We will make special choices of these orderings in Appendix A and assume that from now on all ribbon trees are equipped with the canonical orderings of their internal edges defined in Appendix A.

For any $T \in \mathrm{RTree}_d$, the T-diagonal is constructed in Definition 4.10 as an intersection of the spaces Δ_v, which are oriented manifolds, since M is oriented. Hence, Δ_T is a transverse intersection of oriented manifolds and therefore itself oriented. A delicate issue is that the orientation of Δ_T depends on the order of the intersection of the Δ_v. In Appendix A.4 with an orientation that is independent of this order.

Throughout the rest of this book, we will view the T-diagonals as oriented manifolds with the orientations from Appendix A.4.

Since the space $\mathcal{M}_Y^d(x_0, x_1, \ldots, x_d, T)$ is oriented and the T-diagonal is an oriented submanifold of the oriented manifold $M^{1+2k(T)+d}$ (with the product orientation), the space $\mathcal{A}_Y^d(x_0, x_1, \ldots, x_d, T)$ is an *oriented* manifold in the case of transverse intersection.

Throughout the rest of this book, we will equip the spaces $\mathcal{A}_Y^d(x_0, x_1, \ldots, x_d, T)$ with the orientations from Appendix A.4.

In the course of this chapter, we will further use the following fact from graph theory without giving a proof:

A d-leafed ribbon tree has at most $(d - 2)$ internal edges for every $d \geq 2$. Moreover, a d-leafed ribbon tree has $(d - 2)$ internal edges if and only if it is a binary tree.

6.3   The Higher Order Multiplications and Their Coefficients

We will apply the results from Sect. 6.2 to define homomorphisms whose coefficients are given as twisted oriented intersection numbers of zero-dimensional spaces of perturbed Morse ribbon trees.

Let $x_0, x_1, \ldots, x_d \in \mathrm{Crit} f$ with

$$\mu(x_0) = \sum_{i=1}^d \mu(x_i) + 2 - d .$$

Together with the graph-theoretic fact from the end of Sect. 6.2, Theorem 4.15 implies that for a regular and universal d-perturbation datum $\mathbf{Y}$, the space $\mathcal{A}_{\mathbf{Y}}^d(x_0, x_1, \ldots, x_d, T)$ will be

- a finite set if T is a binary tree,
- empty if T is non-binary.

Thus, the coefficients we are going to introduce will be well-defined.

Definition 6.3 Let $d \in \mathbb{N}$, $d \geq 2$, $\mathbf{Y} = (\mathbf{Y}_T)_{T \in \mathrm{RTree}_d}$ be a regular and universal d-perturbation datum and let $x_0, x_1, \ldots, x_d \in \mathrm{Crit}\, f$ with

$$\mu(x_0) = \sum_{i=1}^{d} \mu(x_i) + 2 - d .$$

We define

$$a_{\mathbf{Y}}^d(x_0, x_1, \ldots, x_d) := (-1)^{\sigma(x_0, x_1, \ldots, x_d)} \sum_{T \in \mathrm{RTree}_d} \#_{\mathrm{alg}} \mathcal{A}_{\mathbf{Y}}^d(x_0, x_1, \ldots, x_d, T) \in \mathbb{Z} ,$$

where

$$\sigma(x_0, x_1, \ldots, x_d) := (n+1)\left(\mu(x_0) + \sum_{i=1}^{d}(d + 1 - i)\mu(x_i)\right) .$$

Here, $\#_{\mathrm{alg}} \mathcal{A}_{\mathbf{Y}}^d(x_0, x_1, \ldots, x_d, T)$ denotes a twisted oriented intersection number of

$$\underline{E}_{\mathbf{Y}_T} : \mathcal{M}_{\mathbf{Y}}^d(x_0, x_1, \ldots, x_d, T) \to M^{1 + 2k(T) + d}$$

with Δ_T for $T \in \mathrm{RTree}_d$, as defined in Appendix A.4. The number $\#_{\mathrm{alg}} \mathcal{A}_{\mathbf{Y}}^d(x_0, x_1, \ldots, x_d, T)$ differs from the oriented intersection number by a sign that only depends on the topological type of T.

Remark 6.4 The sign twist by the parity of $\sigma(x_0, x_1, \ldots, x_d)$ is chosen for technical reasons. More precisely, the sign twists will turn out to make the higher order multiplications we are going to define fulfil the defining equations of an A_∞-algebra. If this correction was omitted, the defining equations would only be satisfied up to signs. Our choice of $\sigma(x_0, x_1, \ldots, x_d)$ is in accordance with Abouzaid's works [Abo09, Abo11] up to notational changes.

We have completed all the necessary preparations for the definition of the higher order multiplications on Morse cochain complexes. Remember that the underlying groups of the Morse chain complex of f are the free abelian groups generated by the critical points of f of corresponding Morse index.

Definition 6.5 Let $d \geq 2$ and $\mathbf{Y} = (\mathbf{Y}_T)_{T \in \mathrm{RTree}_d}$ be a regular and universal d-perturbation datum. Define a map

$$\mu_{d,\mathbf{Y}} : C^*(f)^{\otimes d} \to C^*(f)$$

of degree

$$\deg \mu_{d,\mathbf{Y}} = 2 - d$$

as the $\mathbb{Z}$-linear extension of

$$x_1 \otimes x_2 \otimes \cdots \otimes x_d \mapsto \sum_{\substack{x_0 \in \mathrm{Crit}\, f \\ \mu(x_0) = \sum_{i=1}^d \mu(x_i) + 2 - d}} a_{\mathbf{Y}}^d(x_0, x_1, \ldots, x_d) \cdot x_0$$

for all $x_1, x_2, \ldots, x_d \in \mathrm{Crit}\, f$.

Choosing a regular and universal d-perturbation datum $\mathbf{Y}_d$ for every $d \geq 2$, we thus construct a family of operations

$$\left\{ \mu_{d,\mathbf{Y}_d} : C^*(f)^{\otimes d} \to C^*(f) \right\}_{d \geq 2} .$$

In the remainder of this section, we will show that the Morse cochain complex of f with these multiplications (and with an appropriate choice of an operation $C^*(f) \to C^*(f)$) will be an A_∞-algebra, if we impose a further condition on the family $(\mathbf{Y}_d)_{d \geq 2}$. In particular, it will turn out that the different perturbation data can not be chosen independently of each other, but have to be related in some way. We will make these relations precise in the language of background perturbations.

6.4 One-Dimensional Moduli Spaces and Their Compactifications

We next want to focus on the one-dimensional case of the compactness results from Chap. 5. We will classify the occurring boundary spaces in that case and discuss how to compactify one-dimensional moduli spaces by adding the boundaries.

 Until further mention, we choose and fix $d \geq 2$, a family of critical points $x_0, x_1, \ldots, x_d$ satisfying

$$\mu(x_0) = \sum_{i=1}^d \mu(x_i) + 3 - d \tag{6.3}$$

and a regular and universal d-perturbation datum $\mathbf{Y} = (\mathbf{Y}_T)_{T \in \mathrm{RTree}_d}$.

 We derive from Theorem 4.15 that for $T \in \mathrm{RTree}_d$ the space $\mathcal{A}_{\mathbf{Y}}^d(x_0, x_1, \ldots, x_d, T)$ will be

- a one-dimensional manifold if T is a binary tree,
- a finite set if T has precisely $(d - 3)$ internal edges,
- empty if T has less than $(d - 3)$ internal edges.

We investigate the one-dimensional case in greater detail using the methods of Chap. 5.

Theorem 6.6 *Let* $\mathbf{Y}$ *be a regular and universal perturbation datum and* $T \in \mathrm{RTree}_d$ *be a binary tree. Assume that a sequence in* $\mathcal{A}_{\mathbf{Y}}^d(x_0, x_1, \ldots, x_d, T)$ *does not have a convergent subsequence. Then the sequence will have a geometrically convergent subsequence whose geometric limit lies (up to permutation of its components) in*

$$\bigcup_{e \in E_{int}(T)} \mathcal{A}_{\mathbf{Y}}^d(x_0, x_1, \ldots, x_d, T/e) \cup \bigcup_{\substack{y_0 \in \mathrm{Crit} f \\ \mu(y_0) = \mu(x_0) - 1}} \widehat{\mathcal{M}}(x_0, y_0) \times \mathcal{A}_{\mathbf{Y}}^d(y_0, x_1, \ldots, x_d, T)$$

$$\cup \bigcup_{i=1}^{d} \bigcup_{\substack{y_i \in \mathrm{Crit} f \\ \mu(y_i) = \mu(x_i) + 1}} \mathcal{A}_{\mathbf{Y}}^d(x_0, x_1, \ldots, x_{i-1}, y_i, x_{i+1}, \ldots, x_d, T) \times \widehat{\mathcal{M}}(y_i, x_i) \qquad (6.4)$$

$$\cup \bigcup_{e \in E_{int}(T)} \bigcup_{x_e \in \mathrm{Crit} f} \mathcal{B}_{\mathbf{Y}}\left((x_0, x_e), (x_1, \ldots, x_d, x_e), T, \{e\}\right) . \qquad (6.5)$$

Apart from the one involving a quotient tree, the different types of limit spaces in Theorem 6.6 are illustrated in Fig. 6.1. The left-hand picture shows a space of the form $\widehat{\mathcal{M}}(x_0, y_0) \times \mathcal{A}_{\mathbf{Y}}^d(y_0, x_1, \ldots, x_d, T)$ while the picture in the middle shows a space of type (6.4) and the right-hand one depicts a space of type (6.5).

Proof of Theorem 6.6 From Theorem 5.22 we deduce that any sequence without convergent subsequence has a subsequence which converges geometrically against an element of a product of a space of the form

$$\mathcal{B}_{\mathbf{Y}}\left((y_0, (y_{e-})_{e \in F_1}), (y_1, \ldots, y_d, (y_{e+})_{e \in F_1}), T/F_2, F_1\right)$$

with certain spaces of unparametrized Morse trajectories. By Proposition 5.23, this space is a smooth manifold whose dimension satisfies

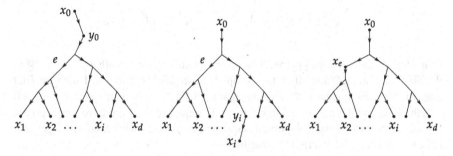

Fig. 6.1 The limit spaces from Theorem 6.6

$$\dim \mathcal{B}_{\mathbf{Y}} \left(\left(y_0, (y_{e-})_{e \in F_1} \right), \left(y_1, \ldots, y_d, (y_{e+})_{e \in F_1} \right), T/F_2, F_1 \right)$$

$$\leq \dim \mathcal{A}_{\mathbf{Y}}^d (x_0, x_1, \ldots, x_d, T) - |F_1| - |F_2| \overset{(6.3)}{\leq} 1 - |F_1| - |F_2| .$$

Therefore, the space can only be nonempty if $|F_1 \cup F_2| \leq 1$, and we have to study the different cases that can occur.

Consider the case of a geometrically convergent subsequence for which $F_1 \neq \emptyset$ in the above notation. This implies that $F_2 = \emptyset$ and that $F_1 = \{e\}$ for some $e \in E_{int}(T)$. One checks without difficulty, using once again Theorem 3.10, that the dimension of any space of the form

$$\mathcal{B}_{\mathbf{Y}} \left((y_0, y_{e-}), (y_1, \ldots, y_d, y_{e+}), T, \{e\} \right)$$

is given by

$$\dim \mathcal{B}_{\mathbf{Y}} \left((y_0, y_{e-}), (y_1, \ldots, y_d, y_{e+}), T, \{e\} \right) = \mu(y_0) - \sum_{q=1}^{d} \mu(y_q) + \mu(e_-) - \mu(e_+) + d - 3 .$$

Inequality (5.15) from Proposition 5.23 implies:

$$\mu(y_0) - \sum_{q=1}^{d} \mu(y_q) + \mu(y_{e-}) - \mu(y_{e+}) + d - 3 \leq \left(\mu(x_0) - \sum_{q=1}^{d} \mu(x_q) + d - 2 \right) - 1$$

$$\Leftrightarrow \mu(y_0) + \sum_{q=1}^{d} \mu(x_q) + \mu(y_{e-}) \leq \mu(x_0) + \sum_{q=1}^{d} \mu(y_q) + \mu(y_{e+}) . \quad (6.6)$$

But by (5.13), this inequality is valid if and only if equality holds and

$$x_0 = y_0 , \quad x_i = y_i \ \forall i \in \{1, 2, \ldots, d\} , \quad y_{e+} = y_{e-} =: x_e .$$

This means that if a sequence in $\mathcal{A}_{\mathbf{Y}}^d (x_0, x_1, \ldots, x_d, T)$ has a geometrically convergent component sequence associated with an internal edge, then the subsequence will converge against an element of

$$\bigcup_{x_e \in \mathrm{Crit} f} \mathcal{B}_{\mathbf{Y}} \left((x_0, x_e), (x_1, \ldots, x_d, x_e), T, \{e\} \right) .$$

Consider the case of a geometrically convergent subsequence with $F_1 = F_2 = \emptyset$ in the above notation. Then all of the component sequences associated with internal edges have subsequences converging against a trajectories with positive interval length. Since we have assumed that the whole sequence does not have a convergent subsequence, there exist component sequences associated with external edges that have geometrically convergent subsequences.

Consequently, the subsequence converges geometrically against a product of un-parametrized Morse trajectories and an element of a space of the form

$$\mathcal{A}_Y^d(y_0, y_1, \ldots, y_d, T)$$

for some $y_0, \ldots, y_d \in \text{Crit } f$ satisfying the conditions from (5.13). Define $I \subset \{0, 1, \ldots, d\}$ by demanding that $y_i \neq x_i$ if and only if $i \in I$. In particular, this means that for every $i \in I$, there is a strict inequality of the form

$$\mu(y_i) > \mu(x_i) \text{ if } i \in \{1, 2, \ldots, d\} \quad \text{and} \quad \mu(y_i) < \mu(x_i) \text{ if } i = 0. \qquad (6.7)$$

The dimension of $\mathcal{A}_Y^d(y_0, y_1, \ldots, y_d, T)$ is then computed as follows:

$$\dim \mathcal{A}_Y^d(y_0, y_1, \ldots, y_d, T) = \mu(y_0) - \sum_{q=1}^d \mu(x_q) + d - 2$$

$$\overset{(6.7)}{\leq} \mu(x_0) - \sum_{q=1}^d \mu(x_q) + d - 2 - |I| = 1 - |I|,$$

by assumption on $x_0, x_1, \ldots, x_d$. Thus, the space $\mathcal{A}_Y^d(y_0, y_1, \ldots, y_d, T)$ is non-empty only if $|I| = 1$, i.e. $I = \{i\}$ for a unique $i \in \{0, 1, \ldots, d\}$.

For $i \in \{1, 2, \ldots, d\}$ this means that the sequence converges geometrically against an element of

$$\mathcal{A}_Y^d(x_0, x_1, \ldots, x_{i-1}, y_{i1}, x_{i+1}, \ldots, x_d, T) \times \prod_{j=1}^{m-1} \widehat{\mathcal{M}}(y_{ij}, y_{i(j+1)}) \times \widehat{\mathcal{M}}(y_{im}, x_i)$$

for some $m \in \mathbb{N}$ and $y_{i1}, \ldots, y_{im} \in \text{Crit } f$, such that

$$\mu(y_{i1}) > \mu(y_{i2}) > \cdots > \mu(y_{im}) > \mu(x_i). \qquad (6.8)$$

But (6.8) obviously implies $\mu(y_{i1}) \leq \mu(x_i) + m$, and therefore

$$\dim \mathcal{A}_Y^d(x_0, x_1, \ldots, x_{i-1}, y_{i1}, x_{i+1}, \ldots, x_d, T)$$

$$= \mu(x_0) - \sum_{q=1}^{i-1} \mu(x_q) - \mu(y_{i1}) - \sum_{q=i+1}^d \mu(x_q) + 2 - d$$

$$\leq \mu(x_0) - \sum_{q=1}^d \mu(x_q) + 2 - d - m = 1 - m,$$

which immediately implies $m = 1$. Hence, the geometric limit we are considering lies in a space of the form

$$\mathcal{A}_{\mathbf{Y}}^d(x_0, x_1, \ldots, x_{i-1}, y_i, x_{i+1}, \ldots, x_d, T) \times \widehat{\mathcal{M}}(y_i, x_i) \, ,$$

for some $y_i \in \mathrm{Crit}\, f$ with $\mu(y_i) = \mu(x_i) + 1$. Analogously, the case $i = 0$ leads to geometric convergence against elements of

$$\widehat{\mathcal{M}}(x_0, y_0) \times \mathcal{A}_{\mathbf{Y}}^d(y_0, x_1, \ldots, x_d, T)$$

for some $y_0 \in \mathrm{Crit}\, f$ with $\mu(y_0) = \mu(x_0) - 1$.

It remains to discuss the case $F_1 = \emptyset$ and $F_2 \neq \emptyset$. Consider a sequence which has a geometrically convergent subsequence of that type and assume for the moment that all component sequences associated with external edges have convergent subsequences. Then the sequence of Morse ribbon trees has a subsequence for which all component subsequences converge. By Theorem 5.15, the projection of that subsequence away from the finite-length trajectories associated with elements of F_2 converges against an element of

$$\mathcal{A}_{\mathbf{Y}}^d(x_0, x_1, \ldots, x_d, T/F_2) \, .$$

This especially requires the space $\mathcal{A}_{\mathbf{Y}}^d(x_0, x_1, \ldots, x_d, T/F_2)$ to be non-empty. Consequently its expected dimension has to be non-negative. But from Theorem 4.15 and the regularity of $\mathbf{Y}$ we know that

$$\dim \mathcal{A}_{\mathbf{Y}}^d(x_0, x_1, \ldots, x_d, T/F_2) = \mu(x_0) - \sum_{i=1}^d \mu(x_i) + k(T) - |F_2|$$

$$= \mu(x_0) - \sum_{i=1}^d \mu(x_i) + d - 2 - |F| \stackrel{(6.3)}{=} 1 - |F_2| \stackrel{!}{\geq} 0 \, .$$

This implies that F contains at most one element. By assumption F is non-empty, so it contains precisely one element, i.e. $F = \{e\}$ for some $e \in E_{int}(T)$. By similar arguments, one argues that for dimensional reasons, the case $F_2 \neq \emptyset$ can not coincide with the geometric convergence of sequences associated with external edges. Hence, there are no further cases to consider.

The claim follows by putting the different cases together.   □

Definition 6.7 Let A be a manifold and B be a compact topological space with $A \subset B$. We say that A *can be compactified to* B if B is homeomorphic to the topological closure of A.

The following theorem requires the use of so-called gluing methods in Morse theory. Analytically, this is a very delicate issue, and we refrain from giving a detailed discussion.

Every situation occurring in the proof requiring gluing analysis is a straightforward application of the standard gluing results, either for perturbed negative or positive semi-infinite or for perturbed finite-length trajectories.

The results for the unperturbed case are stated and proven in [Weh12, Sch99]. See also [KM07, Chap. 18] for the case of finite-length trajectories. The results extend to the perturbed case, since the line of argument used to prove these theorems essentially relies on a local analysis of the moduli spaces of trajectories and locally the perturbed and the unperturbed case are described in the same way.

Remark 6.8 For the first time in this book, we will need condition (2.11) from the definition of the perturbation space $\mathfrak{X}_0(M)$ in the proof of the following statement. It will guarantee the differentiability of a certain map we will need for the boundary description in Theorem 6.9.

Theorem 6.9 *Let $T \in \text{RTree}_d$ be a binary tree. The space $\mathcal{A}_{\mathbf{Y}}^d(x_0, x_1, \ldots, x_d, T)$ can be compactified to a compact one-dimensional manifold $\overline{\mathcal{A}}_{\mathbf{Y}}^d(x_0, x_1, \ldots, x_d, T)$ of class C^{n+1} whose boundary is given by*

$$\partial \overline{\mathcal{A}}_{\mathbf{Y}}^d(x_0, x_1, \ldots, x_d, T) = \bigcup_{\substack{y_0 \in \text{Crit} f \\ \mu(y_0) = \mu(x_0) - 1}} \widehat{\mathcal{M}}(x_0, y_0) \times \mathcal{A}_{\mathbf{Y}}^d(y_0, x_1, \ldots, x_d, T)$$

$$\cup \bigcup_{i=1}^{d} \bigcup_{\substack{y_i \in \text{Crit} f \\ \mu(y_i) = \mu(x_i) + 1}} \mathcal{A}_{\mathbf{Y}}^d(x_0, x_1, \ldots, x_{i-1}, y_i, x_{i+1}, \ldots, x_d, T) \times \widehat{\mathcal{M}}(y_i, x_i)$$

$$\cup \bigcup_{e \in E_{int}(T)} \bigcup_{x_e \in \text{Crit} f} \mathcal{B}_{\mathbf{Y}}\left((x_0, x_e), (x_1, \ldots, x_d, x_e), T, \{e\}\right) \cup \bigcup_{e \in E_{int}(T)} \mathcal{A}_{\mathbf{Y}}^d(x_0, x_1, \ldots, x_d, T/e) .$$

Proof The gluing results from [Weh12, Sch99] and [KM07, Chap. 18] cover three of the four different types of moduli spaces in the boundary. Gluing analysis for perturbed negative semi-infinite trajectories shows that every element of

$$\bigcup_{\substack{y_0 \in \text{Crit} f \\ \mu(y_0) = \mu(x_0) - 1}} \widehat{\mathcal{M}}(x_0, y_0) \times \mathcal{A}_{\mathbf{Y}}^d(y_0, x_1, \ldots, x_d, T)$$

bounds a unique component of $\mathcal{A}_{\mathbf{Y}}^d(x_0, x_1, \ldots, x_d, T)$, while gluing analysis for perturbed positive semi-infinite trajectories shows that every element of

$$\bigcup_{i=1}^{d} \bigcup_{\substack{y_i \in \text{Crit} f \\ \mu(y_i) = \mu(x_i) + 1}} \mathcal{A}_{\mathbf{Y}}^d(x_0, x_1, \ldots, x_{i-1}, y_i, x_{i+1}, \ldots, x_d, T) \times \widehat{\mathcal{M}}(y_i, x_i)$$

bounds a unique component of $\mathcal{A}_{\mathbf{Y}}^d(x_0, x_1, \ldots, x_d, T)$. Furthermore, gluing analysis for perturbed finite-length trajectories shows that indeed every element of

$$\bigcup_{e \in E_{int}(T)} \bigcup_{x_e \in \text{Crit} f} \mathcal{B}_{\mathbf{Y}}\left((x_0, x_e), (x_1, \ldots, x_d, x_e), T, \{e\}\right)$$

bounds a unique component of $\mathcal{A}_{\mathbf{Y}}^d(x_0, x_1, \ldots, x_d, T)$.

The proof that $\bigcup_{e \in E_{int}(T)} \mathcal{A}_Y^d(x_0, x_1, \ldots, x_d, T/e)$ is indeed contained in the boundary of $\mathcal{A}_Y^d(x_0, x_1, \ldots, x_d, T)$ requires the use of a slightly different method, and we will perform this proof in greater detail. Its main idea and structure will be the following for a given $e \in E_{int}(T)$:

- Instead of considering families of trajectories associated with *all* edges of T and intersecting it transversely with Δ_T, we consider certain families of trajectories associated with *all edges but e*.
- In reverse, we will not use the map $\underline{E}_T$ to intersect the family with Δ_T, but another map to be defined, which "includes" the intersection conditions involving e. (We will make this precise in the course of the proof.)
- While all edge lengths of elements in $\mathcal{A}_Y^d(x_0, x_1, \ldots, x_d, T)$ are by definition positive, the reformulated transversality problem will naturally extend to the case of the trajectory associated with e having length zero.
- We show that in the case of a regular perturbation datum, the evaluation map on the families of trajectories under discussion will intersect both the interior and the boundary of this submanifold transversely.
- We use the universality of the perturbation datum to identify the preimage of Δ_T under the abovementioned map with

$$\mathcal{A}_Y^d(x_0, x_1, \ldots, x_d, T) \cup \mathcal{A}_Y^d(x_0, x_1, \ldots, x_d, T/e)$$

to show that this space is a one-dimensional manifold with boundary.

Let $e \in E_{int}(T)$ be fixed throughout the discussion and put $k := k(T)$. Consider the following map whose domain is a smooth manifold with boundary:

$$g : [0, +\infty) \times (0, +\infty)^{k-1} \times M \to M , \quad \left(l, \vec{l}, x\right) \mapsto \phi_{0,l}^{Y_e(\vec{l})}(x) .$$

Here, $\phi_{0,l}^{Y_e(\vec{l})}$ denotes the time-l map of the time-dependent vector field $Y_e\left(\vec{l}\right)$ with respect to the initial time zero. In other words, it is the time-l-map of the vector field $Y_e\left(\vec{l}, 0, \cdot\right) : M \to TM$. In particular,

$$g\left(0, \vec{l}, x\right) = x \tag{6.9}$$

for every $\vec{l} \in (0, +\infty)^{k-1}, x \in M$. Hence, the restriction of g to $\{0\} \times (0, +\infty)^{k-1} \times M$ is independent of the $(0, +\infty)^{k-1}$-component. Condition (2.11) in the definition of $\mathfrak{X}_0(M)$ implies that the vector field Y_e and all existing derivatives tend to zero for $l \to 0$, so g is of class C^{n+1} at every $(0, \vec{l}, x)$. The Parametrized Flow Theorem, see [AR67, Theorem 21.4], implies that map g is of class C^{n+1} away from $l = 0$, hence g is of class C^{n+1}.

Throughout the rest of the proof, we assume w.l.o.g. that e is the *first* edge in $E_{int}(T)$ according to the chosen ordering of $E_{int}(T)$. To prepare the following, we introduce another space whose meaning will become apparent momentarily:

$$\widetilde{\mathcal{M}}_{\mathbf{Y}}^{d}(T) := \big\{ (l_e, \gamma_0, (l_f, \gamma_f)_{f \neq e}, \gamma_1, \ldots, \gamma_d) \mid l_e \in [0, +\infty), \ \gamma_0 \in W^{-}\left(x_0, Y_0((l_f)_{f \neq e}, l_e)\right),$$
$$(l_h, \gamma_h) \in \mathcal{M}\left(Y_h\left((l_f)_{f \notin \{e,h\}}, l_e\right)\right) \ \forall h \in E_{int}(T) \setminus \{e\},$$
$$\gamma_i \in W^{+}\left(x_i, Y_i\left((l_f)_{f \neq e}, l_e\right)\right) \ \forall i \in \{1, 2, \ldots, d\}\big\} .$$

One checks without difficulties that $\widetilde{\mathcal{M}}_{\mathbf{Y}}^{d}(T)$ is a manifold with boundary of class C^{n+1} which is diffeomorphic to

$$[0, +\infty) \times \mathcal{W}^{u}(x_0) \times \prod_{f \in E_{int}(T) \setminus \{e\}} \mathcal{M} \times \mathcal{W}^{s}(x_1) \times \cdots \times \mathcal{W}^{s}(x_d)$$

and that the following map is a submersion of class C^{n+1}:

$$\mathcal{M}_{\mathbf{Y}}^{d}(x_0, x_1, \ldots, x_d, T) \to \widetilde{\mathcal{M}}_{\mathbf{Y}}^{d}(T) ,$$
$$\left(\gamma_0, (l_f, \gamma_f)_{f \in E_{int}(T)}, \gamma_1, \ldots, \gamma_d\right) \mapsto \left(l_e, \gamma_0, (l_f, \gamma_f)_{f \neq e}, \gamma_1, \ldots, \gamma_d\right) . \tag{6.10}$$

Let $\left(\gamma_0, (l_e, \gamma_e), (l_f, \gamma_f)_{f \neq e}, \gamma_1, \ldots, \gamma_d\right) \in \mathcal{M}_{\mathbf{Y}}^{d}(x_0, x_1, \ldots, x_d, T)$. This particularly means that

$$(l_e, \gamma_e) \in \mathcal{M}\left(Y_e\left((l_f)_{f \neq e}\right)\right) .$$

In terms of the map g, this is equivalent to

$$\gamma_e(s) = \phi_{0,s}^{Y_e((l_f)_{f \neq e})}(\gamma_e(0)) = g\left(s, (l_f)_{f \neq e}, \gamma_e(0)\right) \quad \forall s \in [0, l_e] . \tag{6.11}$$

By definition, $\left(\gamma_0, (l_e, \gamma_e), (l_f, \gamma_f)_{f \neq e}, \gamma_1, \ldots, \gamma_d\right) \in \mathcal{A}_{\mathbf{Y}}^{d}(x_0, x_1, \ldots, x_d, T)$ if and only if

$$\underline{E}_T\left(\gamma_0, (l_e, \gamma_e), (l_f, \gamma_f)_{f \neq e}, \gamma_1, \ldots, \gamma_d\right) \in \Delta_T$$
$$\Leftrightarrow \left(\gamma_0(0), (\gamma_e(0), \gamma_e(l_e)), (\gamma_f(0), \gamma_f(l_f))_{f \neq e}, \gamma_1(0), \ldots, \gamma_d(0)\right) \in \Delta_T$$
$$\overset{(6.11)}{\Leftrightarrow} \left(\gamma_0(0), (\gamma_e(0), g\left(l_e, (l_f)_{f \neq e}, \gamma_e(0)\right)), (\gamma_f(0), \gamma_f(l_f))_{f \neq e}, \gamma_1(0), \ldots, \gamma_d(0)\right) \in \Delta_T .$$

Let $e' \in E(T)$ be the edge with $v_{out}(e') = v_{in}(e)$. We assume w.l.o.g. that $e' \in E_{int}(T)$, the remaining case $e' = e_0(T)$ is discussed along the same lines.

Using the above equivalence, one checks that the map from (6.10) maps the space $\mathcal{A}_{\mathbf{Y}}^{d}(x_0, x_1, \ldots, x_d, T)$ homeomorphically onto the following space:

$$\widetilde{\mathcal{A}}_{\mathbf{Y}}^{d}(T) := \Big\{ (l_e, \gamma_0, (l_f, \gamma_f)_{f \neq e}, \gamma_1, \ldots, \gamma_d) \in \widetilde{\mathcal{M}}_{\mathbf{Y}}^{d}(T) \mid l_e > 0,$$
$$(\gamma_{e'}(l_{e'}), g\left(l_e, (l_f)_{f \neq e}, \gamma_{e'}(l_{e'})\right)), (\gamma_0(0), (\gamma_f(0), \gamma_f(l_f))_{f \neq e}, \gamma_1(0), \ldots, \gamma_d(0)) \in \Delta_T\Big\} .$$

We will reformulate $\mathcal{A}_{\mathbf{Y}}^d(x_0, x_1, \ldots, x_d, T/e)$ in a very similar way. Without further mentioning, we will identify $E_{int}(T/e)$ with $E_{int}(T) \setminus \{e\}$.

Let $\left(\gamma_0, (l_f, \gamma_f)_{f \neq e}, \gamma_1, \ldots, \gamma_d\right) \in \mathcal{M}_{\mathbf{Y}}^d(x_0, x_1, \ldots, x_d, T/e)$. Then

$$\left(\gamma_0, (l_f, \gamma_f)_{f \neq e}, \gamma_1, \ldots, \gamma_d\right) \in \mathcal{A}_{\mathbf{Y}}^d(x_0, x_1, \ldots, x_d, T/e)$$

$$\Leftrightarrow \; \underline{E}_{T/e}\left(\gamma_0, (l_f, \gamma_f)_{f \neq e}, \gamma_1, \ldots, \gamma_d\right) \in \Delta_{T/e}$$

$$\Leftrightarrow \; \left(\gamma_0(0), (\gamma_f(0), \gamma_f(l_f))_{f \neq e}, \gamma_1(0), \ldots, \gamma_d(0)\right) \in \Delta_{T/e} \, .$$

Comparing the definitions of Δ_T and $\Delta_{T/e}$ and letting e' be given as above, this condition is equivalent to

$$\left(\gamma_0(0), (\gamma_{e'}(0), \gamma_{e'}(0)), (\gamma_f(0), \gamma_f(l_f))_{f \neq e}, \gamma_1(0), \ldots, \gamma_d(0)\right) \in \Delta_T$$

$$\overset{(6.9)}{\Leftrightarrow} \; \left(\gamma_0(0), \left(\gamma_{e'}(0), g\left(0, (l_f)_{f \neq e}, \gamma_{e'}(0)\right)\right), (\gamma_f(0), \gamma_f(l_f))_{f \neq e}, \gamma_1(0), \ldots, \gamma_d(0)\right) \in \Delta_T \, .$$

Using this condition and the universality of the perturbation datum, one checks that $\mathcal{A}_{\mathbf{Y}}^d(x_0, x_1, \ldots, x_d, T/e)$ is homeomorphic to the following space:

$$\widetilde{\mathcal{A}}_{\mathbf{Y}}^d(T/e) := \Big\{ (0, \gamma_0, (l_f, \gamma_f)_{f \neq e}, \gamma_1, \ldots, \gamma_d) \, \Big| \, \gamma_0 \in W^-\left(x_0, Y_0((l_f)_{f \neq e}, 0)\right) \, ,$$

$$(l_h, \gamma_h) \in \mathcal{M}\left(Y_h\left((l_f)_{f \notin \{e, h\}}, 0\right)\right) \; \forall h \in E_{int}(T) \setminus \{e\} \, ,$$

$$\gamma_i \in W^+\left(x_i, Y_i\left((l_f)_{f \neq e}, 0\right)\right) \; \forall i \in \{1, 2, \ldots, d\} \, ,$$

$$\left(\gamma_0(0), \left(\gamma_{e'}(0), g\left(0, (l_f)_{f \neq e}, \gamma_{e'}(0)\right)\right), (\gamma_f(0), \gamma_f(l_f))_{f \neq e}, \gamma_1(0), \ldots, \gamma_d(0)\right) \in \Delta_T \Big\}$$

$$= \Big\{ (0, \gamma_0, (l_f, \gamma_f)_{f \neq e}, \gamma_1, \ldots, \gamma_d) \in \widetilde{\mathcal{M}}_{\mathbf{Y}}^d$$

$$\Big| \left(\gamma_0(0), \left(\gamma_{e'}(0), g\left(0, (l_f)_{f \neq e}, \gamma_{e'}(0)\right)\right), (\gamma_f(0), \gamma_f(l_f))_{f \neq e}, \gamma_1(0), \ldots, \gamma_d(0)\right) \in \Delta_T \Big\} \, .$$

Note the strong similarity between the spaces $\widetilde{\mathcal{A}}_{\mathbf{Y}}^d(T)$ and $\widetilde{\mathcal{A}}_{\mathbf{Y}}^d(T/e)$. To state a transversality problem including both spaces, we consider the map

$$\widetilde{E} : \widetilde{\mathcal{M}}_{\mathbf{Y}}^d(T) \to M^{1+2k+d} \, , \quad \left(l_e, \gamma_0, (l_f, \gamma_f)_{f \neq e}, \gamma_1, \ldots, \gamma_d\right) \mapsto$$

$$\left(\gamma_0(0), (\gamma_f(0), \gamma_f(l_f))_{f \neq e}, \left(\gamma_{e'}(l_{e'}), g\left(l_e, (l_f)_{f \neq e}, \gamma_{e'}(l_{e'})\right)\right), \gamma_1(0), \ldots, \gamma_d(0)\right) \, .$$

Taking a close look at the definitions of $\widetilde{\mathcal{A}}_{\mathbf{Y}}^d(T)$ and $\widetilde{\mathcal{A}}_{\mathbf{Y}}^d(T/e)$, one deduces that

$$\widetilde{E}^{-1}(\Delta_T) = \widetilde{\mathcal{A}}_{\mathbf{Y}}^d(T) \cup \widetilde{\mathcal{A}}_{\mathbf{Y}}^d(T/e) \, .$$

The regularity of $\mathbf{Y}$ implies that both $\widetilde{E}$ and its restriction to the boundary of $\widetilde{\mathcal{M}}_{\mathbf{Y}}^d(T)$ are transverse to Δ_T. Hence, $\widetilde{E}^{-1}(\Delta_T)$ is a one-dimensional manifold with boundary of class C^{n+1}. Its boundary is given by $\widetilde{\mathcal{A}}_{\mathbf{Y}}^d(x_0, x_1, \ldots, x_d, T/e)$ while its interior is given by $\widetilde{\mathcal{A}}_{\mathbf{Y}}^d(x_0, x_1, \ldots, x_d, T)$. In the light of the above identifications, this shows that every element of $\mathcal{A}_{\mathbf{Y}}^d(x_0, x_1, \ldots, x_d, T/e)$ bounds a unique component of $\mathcal{A}_{\mathbf{Y}}^d(x_0, x_1, \ldots, x_d, T)$, which we had to show. $\square$

6.5 Explicit Descriptions of Boundary Spaces

The significance of Theorem 6.9 will become apparent after a deeper investigation of the spaces $\mathcal{B}_\mathbf{Y}((x_0, x_e), (x_1, \ldots, x_d, x_e), T, \{e\})$. It will turn out that, if the d-perturbation datum $\mathbf{Y}$ fulfills some additional condition, then for every $e \in E_{int}(T)$ and every choice of $x_e \in \operatorname{Crit} f$ the space $\mathcal{B}_\mathbf{Y}((x_0, x_e), (x_1, \ldots, x_d, x_e), T, \{e\})$ will be diffeomorphic to a product of two moduli spaces of perturbed Morse ribbon trees. This will be the key observation for proving that the defining equations of an A_∞-algebra are satisfied. The latter will be introduced in Sect. 6.7.

We will use the following graph-theoretic lemma without proof:

Lemma 6.10 *Let $T \in \mathrm{RTree}_d$, $e \in E_{int}(T)$. There are unique numbers $i \in \{1, 2, \ldots, d\}$ and $l \in \{0, 1, \ldots, d-i\}$ such that*

$$e \in \bigcap_{j=1}^{i-1} E(P_j(T))^c \cap \bigcap_{k=i}^{i+l} E(P_k(T)) \cap \bigcap_{m=i+l+1}^{d} E(P_m(T))^c$$

where $\cdot^c$ denotes the complement in $E(T)$ and where $P_j(T)$ denotes the path from the root of T to its jth leaf for every $j \in \{1, 2, \ldots, d\}$.

See Fig. 6.2 for an illustration of the situation in Lemma 6.10.

Definition 6.11 Let $T \in \mathrm{RTree}_d$, $e \in E_{int}(T)$. If i and l are the numbers associated to e by Lemma 6.10, then we say that e *is of type* (i, l) in T.

Consider a fixed $e \in E_{int}(T)$. We can construct a tree T^e out of T by removing e from the edges of T, inserting a new vertex v_e and two new edges f_1, f_2, such that:

$$V(T^e) = V(T) \cup \{v_e\}, \qquad E(T^e) = (E(T) \setminus \{e\}) \cup \{f_1, f_2\},$$
$$v_{\mathrm{out}}(f_1) = v_e, \quad v_{\mathrm{in}}(f_1) = v_{\mathrm{out}}(f) \;\; \forall f \in E_{int}(T) \text{ with } v_{\mathrm{out}}(f) = v_{\mathrm{in}}(e),$$
$$v_{\mathrm{in}}(f_2) = v_e, \quad v_{\mathrm{out}}(f_2) = v_{\mathrm{in}}(f) \;\; \forall f \in E_{int}(T) \text{ with } v_{\mathrm{in}}(f) = v_{\mathrm{out}}(e).$$

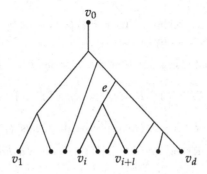

Fig. 6.2 e is of type (i, l) in $T \in \mathrm{RTree}_d$

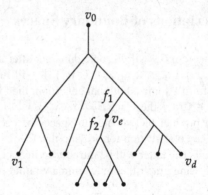

Fig. 6.3 The tree T^e for T and e from Fig. 6.2

For the tree T and the edge e from Fig. 6.2, the tree T^e is depicted in Fig. 6.3. Note that T^e is not a ribbon tree, since the new vertex v_e is by definition binary. Nevertheless, we can always decompose the tree T^e into two ribbon trees as follows.

Definition 6.12 Let $e \in E_{int}(T)$ be of type (i, l) for some numbers $i \in \{1, 2, \ldots, d\}$ and $l \in \{1, 2, \ldots, d - i\}$. We define two subtrees $T_1^e, T_2^e \subset T^e$ with

$$\left(T_1^e, T_2^e\right) \in \text{RTree}_{d-l} \times \text{RTree}_{l+1}$$

by demanding that
$$v_0\left(T_1^e\right) = v_0(T) , \quad v_0\left(T_2^e\right) = v_e ,$$

that the leaves of T_1^e are given by

$$\{v_1(T), \ldots, v_{i-1}(T), v_e, v_{i+l+1}(T), \ldots, v_d(T)\}$$

and that the leaves of T_2^e are given by

$$\{v_i(T), v_{i+1}(T), \ldots, v_{i+l}(T)\} .$$

If we demand all of these conditions, the pair (T_1^e, T_2^e) will be well-defined. We call (T_1^e, T_2^e) *the splitting of T along e.*

The situation is depicted in Fig. 6.4. Note that especially:

$$E_{int}(T_1^e) \cup E_{int}(T_2^e) = E_{int}(T) \setminus \{e\} .$$

We state an important property of the respective diagonals.

Lemma 6.13 *Let $T \in \text{RTree}_d$ and $e \in E_{int}(T)$. There is a diffeomorphism*

$$\Delta_T \xrightarrow{\cong} \Delta_{T_1^e} \times \Delta_{T_2^e} .$$

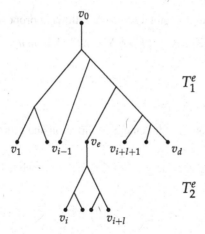

Fig. 6.4 The splitting of T into (T_1^e, T_2^e) for the example from Fig. 6.2

Proof Consider the following permutation:

$$M^{1+2k(T)+d} \overset{\cong}{\to} M^{1+2k(T_1^e)+d-l} \times M^{1+2k(T_2^e)+l+1}$$

$$\left(q_0, \left(q_{in}^f, q_{out}^f\right)_{f\in E_{int}(T)}, q_1, \ldots, q_d\right) \mapsto$$

$$\left(q_0, \left(q_{in}^f, q_{out}^f\right)_{f\in E_{int}(T_1^e)}, q_1, \ldots q_{i-1}, q_{in}^e, q_{i+l+1}, \ldots, q_d,\right.$$

$$\left. q_{out}^e, \left(q_{in}^f, q_{out}^f\right)_{f\in E_{int}(T_2^e)}, q_i, \ldots, q_{i+l}\right),$$

where the components associated with internal edges are reordered according to the chosen orderings of $E_{int}(T_1^e)$ and $E_{int}(T_2^e)$. Comparing the definitions of the respective diagonals, one checks that this permutation induces the desired diffeomorphism.
□

Let $T \in \text{RTree}_d$ be a binary tree and $e \in E_{int}(T)$. Before we continue, we are going to make a few remarks on the space $\mathcal{B}_Y\left((x_0, x_e), (x_1, \ldots, x_d, x_e), T, \{e\}\right)$. In Sect. 4.3 we have written background perturbations $\mathbf{X} \in \mathfrak{X}^{back}(T)$ as

$$\mathbf{X} = \left(\mathbf{X}^-, \left(\mathbf{X}_e^0\right)_{e\in E_{int}(T)}, \mathbf{X}_1^+, \ldots, \mathbf{X}_d^+\right), \quad \text{where}$$

$$\mathbf{X}^- = \left(X_e^-\right)_{e\in E_{int}(T)}, \quad \mathbf{X}_e^0 = \left(\left(X_{ef}^0\right)_{f\in E_{int}(T)\setminus\{e\}}, X_{e+}, X_{e-}\right) \forall e \in E_{int}(T),$$

$$\mathbf{X}_i^+ = \left(X_{ei}^+\right)_{e\in E_{int}(T)} \quad \forall i \in \{1, 2, \ldots, d\}.$$

In this section, we prefer to reorder the components of $\mathbf{X}$ and to write:

$$\mathbf{X} = \left(X_e^-, (X_{ef}^0)_{f\in E_{int}(T)\setminus\{e\}}, X_{e+}, X_{e-}, X_{e1}^+, \ldots, X_{ed}^+\right)_{e\in E_{int}(T)} =: (\mathbf{X}_e)_{e\in E_{int}(T)}.$$
$$(6.12)$$

The reason for this change of notation is the following proposition.

Proposition 6.14 *Let* $\mathbf{X} \in \mathfrak{X}^{back}(T)$. *If* $\mathbf{Y} \in \mathfrak{X}(T, \mathbf{X})$, *then*

$$\mathbf{Y}_{\{e\}} = \mathbf{X}_e$$

for every $e \in E_{int}(T)$ *in the notation of (5.8).*

Proof This is nothing but a slight change of notation and follows immediately from writing down the definitions.   □

By definition of $\mathcal{B}_\mathbf{Y} \left((x_0, x_e), (x_1, \ldots, x_d, x_e), T, \{e\} \right)$, we know that for every element

$$\left(\gamma_0, \left(l_f, \gamma_f \right)_{f \in E_{int}(T) \backslash e}, \gamma_+, \gamma_-, \gamma_1, \ldots, \gamma_d \right) \in \mathcal{B}_\mathbf{Y} \left((x_0, x_e), (x_1, \ldots, x_d, x_e), T, \{e\} \right)$$

the following holds:

$$\left(\gamma_0(0), \left(\gamma_f(0), \gamma_f(l_f) \right)_{f \in E_{int}(T) \backslash \{e\}}, \gamma_+(0), \gamma_-(0), \gamma_1(0), \ldots, \gamma_d(0) \right) \in \sigma_{\{e\}}(\Delta_T) \,,$$

where $\sigma_{\{e\}}$ is given by the map σ_F from Theorem 5.17 with $F = \{e\}$. By Lemma 6.13 and the permutation constructed in its proof, the above is equivalent to

$$\left(\left(\gamma_0(0), \left(\gamma_f(0), \gamma_f(l_f) \right)_{f \in E_{int}(T_1^e)}, \gamma_1(0), \ldots, \gamma_{i-1}(0), \gamma_+(0), \gamma_{i+l+1}(0), \ldots, \gamma_d(0) \right), \right.$$
$$\left. \left(\gamma_-(0), \left(\gamma_f(0), \gamma_f(l_f) \right)_{f \in E_{int}(T_2^e)}, \gamma_i(0), \ldots, \gamma_{i+l}(0) \right) \right) \in \Delta_{T_1^e} \times \Delta_{T_2^e} \,,$$

or equivalently,

$$\left(\gamma_0(0), \left(\gamma_f(0), \gamma_f(l_f) \right)_{f \in E_{int}(T_1^e)}, \gamma_1(0), \ldots, \gamma_{i-1}(0), \gamma_+(0), \ldots, \gamma_d(0) \right) \in \Delta_{T_1^e} \,, \tag{6.13}$$

$$\left(\gamma_-(0), \left(\gamma_f(0), \gamma_f(l_f) \right)_{f \in E_{int}(T_2^e)}, \gamma_i(0), \ldots, \gamma_{i+l}(0) \right) \in \Delta_{T_2^e} \,. \tag{6.14}$$

Moreover, if $\mathbf{Y} \in \mathfrak{X}(T, \mathbf{X})$ for some $\mathbf{X} \in \mathfrak{X}^{back}(T)$ as in (6.12), Proposition 6.14 implies

$$\gamma_0 \in W^- \left(x_0, X_{e0} \left((l_f)_{f \in E_{int}(T) \backslash \{e\}} \right) \right) \,, \tag{6.15}$$

$$(l_f, \gamma_f) \in \mathcal{M} \left(X_{ef} \left((l_g)_{g \in E_{int}(T) \backslash \{e, f\}} \right) \right) \quad \forall f \in E_{int}(T) \backslash \{e\}, \tag{6.16}$$

$$\gamma_i \in W^+ \left(x_i, X_{ei} \left((l_f)_{f \in E_{int}(T) \backslash \{e\}} \right) \right) \quad \forall i \in \{1, 2, \ldots, d\} \,, \tag{6.17}$$

$$\gamma_+ \in W^+ \left(x_e, X_{e+} \left((l_f)_{f \in E_{int}(T) \backslash \{e\}} \right) \right), \quad \gamma_- \in W^- \left(x_e, X_{e-} \left((l_f)_{f \in E_{int}(T) \backslash \{e\}} \right) \right) \,. \tag{6.18}$$

Comparing these properties with (6.13) and (6.14), it seems plausible to hope that for a convenient choice of $\mathbf{X}$, we can reorder the components of

$$\underline{\gamma} := \left(\gamma_0, \left(l_f, \gamma_f \right)_{f \in E_{int}(T) \setminus \{e\}}, \gamma_+, \gamma_-, \gamma_1, \dots, \gamma_d \right)$$

to obtain an element of the product space

$$\mathcal{A}_{\mathbf{Y}_1}^{d-l}(x_0, x_1, \dots, x_{i-1}, x_e, x_{i+l+1}, \dots, x_d, T_1^e) \times \mathcal{A}_{\mathbf{Y}_2}^{l+1}(x_i, \dots, x_{i+l}, T_2^e) \quad (6.19)$$

for certain perturbation data $\mathbf{Y}_1 \in \mathfrak{X}(T_1^e)$ and $\mathbf{Y}_2 \in \mathfrak{X}(T_2^e)$. This does not hold true for arbitrary choices of $\mathbf{X}$, since the perturbing vector field associated with a component of γ depends on the parameters l_f for *all* $f \in E_{int}(T) \setminus \{e\}$. A necessary and sufficient condition for an identification of

$$\mathcal{B}_{\mathbf{Y}}\left((x_0, x_e), (x_1, \dots, x_d, x_e), T, \{e\} \right)$$

with a product of moduli spaces as in (6.19) is that the parametrized vector fields in $\mathbf{X}_e$ associated with an edge of T_i^e, $i \in \{1, 2\}$, depend *only on those parameters l_f with $f \in E(T_i^e)$*.

In the following, we will introduce a method of constructing background perturbations. Given a k-perturbation datum for every $k < d$, we will inductively construct a background perturbation having the desired property for every d-leafed tree.

Assume we have chosen a family $\mathbf{Y} = \left(\mathbf{Y}^k \right)_{2 \le k < d}$ of perturbations, where $\mathbf{Y}^k$ is a k-perturbation datum for every $2 \le k < d$. We can write $\mathbf{Y}$ as

$$\mathbf{Y} = (\mathbf{Y}_T)_{T \in \bigcup_{k=2}^{d-1} \text{RTree}_k} \, .$$

For every $T \in \text{RTree}_d$ we will construct an $\mathbf{X}^{\mathbf{Y}} \in \mathfrak{X}^{\text{back}}(T)$ out of the family $\mathbf{Y}$. Fix $T \in \text{RTree}_d$ and let $e \in E_{int}(T)$. We have seen above that

$$k(T_1^e) = d - l < d, \quad k(T_2^e) = l + 1 < d \, .$$

Therefore, the family $\mathbf{Y}$ especially contains perturbation data for the trees T_1^e and T_2^e which we denote by

$$\mathbf{Y}_{T_1^e} = \left(Y_{10}, \left(Y_{1f} \right)_{f \in E_{int}(T_1^e)}, Y_{11}, \dots, Y_{1(d-l)} \right) ,$$

$$\mathbf{Y}_{T_2^e} = \left(Y_{20}, \left(Y_{2f} \right)_{f \in E_{int}(T_2^e)}, Y_{21}, \dots, Y_{2(l+1)} \right) .$$

If e is of type (i, l) we define $\mathbf{X}^{\mathbf{Y}} = \left(\mathbf{X}_e^{\mathbf{Y}} \right)_{e \in E_{int}(T)} \in \mathfrak{X}^{\text{back}}(T)$ by

$$\mathbf{X}_e^{\mathbf{Y}} := \left(Y_{10}, \left(Z_f \right)_{f \in E_{int}(T) \setminus \{e\}}, (Y_{1i}, Y_{20}), \right.$$
$$\left. Y_{11}, \dots, Y_{1(i-1)}, Y_{21}, \dots, Y_{2(l+1)}, Y_{1(i+1)}, \dots, Y_{1(d-l)} \right) , \quad (6.20)$$

where

$$Z_f := \begin{cases} Y_{1f} & \text{if } f \in E_{int}\left(T_1^e\right), \\ Y_{2f} & \text{if } f \in E_{int}\left(T_2^e\right). \end{cases}$$

Here, we identify every

$$Y_{1j} \in \mathfrak{X}_{\pm}(M, k(T_1^e)), \ j \in \{0, 1, \ldots, d - l\},$$

with the corresponding element of $\mathfrak{X}_{\pm}(M, k(T))$ which is independent of all parameters associated with internal edges which do not belong to $E_{int}(T_1^e)$. Analogously, we identify every $Y_{2i} \in \mathfrak{X}_{\pm}(M, k(T_2^e))$, $i \in \{0, 1, \ldots, l + 1\}$, with the corresponding element of $\mathfrak{X}_{\pm}(M, k(T))$ which is independent of all parameters associated with internal edges which do not belong to $E_{int}(T_2^e)$. Similar identifications hold for the perturbations Y_{1f} and Y_{2g} for all $f \in E_{int}(T_1^e)$ and $g \in E_{int}(T_2^e)$.

Definition 6.15 (a) A family

$$\mathbf{Y} = (\mathbf{Y}_d)_{d \geq 2},$$

where $\mathbf{Y}_d$ is a d-perturbation datum for every $d \geq 2$, is called a *perturbation datum*. $\mathbf{Y}$ will be called *regular* if $\mathbf{Y}_d$ is regular for every $d \geq 2$. $\mathbf{Y}$ will be called *universal* if $\mathbf{Y}_d$ is universal for every $d \geq 2$.

(b) A perturbation datum $\mathbf{Y} = (\mathbf{Y}_d)_{d \geq 2}$ will be called *consistent* if for every $d \geq 2$ and $T \in \text{RTree}_d$ the following holds:

$$\mathbf{Y}_T \in \mathfrak{X}\left(T, \mathbf{X}^{\mathbf{Y}_{<d}}\right), \tag{6.21}$$

where $\mathbf{Y}_{<d} := (\mathbf{Y}_k)_{2 \leq k < d}$ and $\mathbf{X}^{\mathbf{Y}_{<d}} \in \mathfrak{X}^{\text{back}}(T)$ is the background perturbation associated with $\mathbf{Y}_{<d}$ as described in (6.20).

(c) A perturbation datum is called *admissible* if it is regular, universal and consistent.

If $\mathbf{Y} = (\mathbf{Y}_d)_{d \geq 2}$ is a regular and universal perturbation datum we will write

$$a_{\mathbf{Y}}^d(x_0, x_1, \ldots, x_d) := a_{\mathbf{Y}_d}^d(x_0, x_1, \ldots, x_d)$$

for the twisted intersection numbers from Definition 6.3 and $\mu_{d,\mathbf{Y}} := \mu_{d,\mathbf{Y}_d}$ for the maps from Definition 6.5.

Our introduction of admissible perturbation data is justified by the following lemma.

Lemma 6.16 *Admissible perturbation data exist.*

Proof We will argue inductively over the number of leaves of the trees.

For $d = 2$, there are no consistency conditions, so we can use an arbitrary regular and universal 2-perturbation datum $\mathbf{Y}_2$ to build an admissible perturbation datum. Assume now that for some $d \geq 2$ we have found a family $(\mathbf{Y}_k)_{2 \leq k < d}$, where $\mathbf{Y}_k$ is a k-perturbation datum for every k, such that the consistency condition (6.21) is satisfied for every ribbon tree with at most $d - 1$ leaves.

Let $\mathbf{X}^\mathbf{Y} = (\mathbf{X}^\mathbf{Y}_T)_{T \in \mathrm{RTree}_d}$ be the background d-perturbation datum constructed out of the family $(\mathbf{Y}_k)_{2 \le k < d}$ as described above. Since $\mathbf{Y}_k$ is universal for every $k < d$, it follows that $\mathbf{X}^\mathbf{Y}$ is universal. Thus, by Lemma 6.1 there exists a regular and universal d-perturbation datum $\mathbf{Y}_d = (\mathbf{Y}_T)_{T \in \mathrm{RTree}_d}$, such that $\mathbf{Y}_T \in \mathfrak{X}(T, \mathbf{X}^\mathbf{Y}_T)$ for every $T \in \mathrm{RTree}_d$. By definition of $\mathbf{X}^\mathbf{Y}$, this d perturbation datum is consistent, hence admissible. $\qquad\square$

Theorem 6.17 *Let $\mathbf{Y}$ be an admissible perturbation datum, let $d \ge 2$, $T \in \mathrm{RTree}_d$ and $x_0, x_1, \ldots, x_d \in \mathrm{Crit}\, f$. Then $\mathcal{A}^d_\mathbf{Y}(x_0, x_1, \ldots, x_d, T)$ can be compactified to a compact one-dimensional manifold with boundary $\overline{\mathcal{A}}^d_\mathbf{Y}(x_0, x_1, \ldots, x_d, T)$, whose boundary is given by*

$$\partial \overline{\mathcal{A}}^d_\mathbf{Y}(x_0, x_1, \ldots, x_d, T) = \bigcup_{e \in E_{int}(T)} \mathcal{A}^d_\mathbf{Y}(x_0, x_1, \ldots, x_d, T/e)$$

$$\cup \bigcup_{\substack{y_0 \in \mathrm{Crit}\, f \\ \mu(y_0) = \mu(x_0) - 1}} \widehat{\mathcal{M}}(x_0, y_0) \times \mathcal{A}^d_\mathbf{Y}(y_0, x_1, \ldots, x_d, T)$$

$$\cup \bigcup_{i=1}^d \bigcup_{\substack{y_i \in \mathrm{Crit}\, f \\ \mu(y_i) = \mu(x_i) + 1}} \mathcal{A}^d_\mathbf{Y}(x_0, x_1, \ldots, x_{i-1}, y_i, x_{i+1}, \ldots, x_d, T) \times \widehat{\mathcal{M}}(y_i, x_i)$$

$$\cup \bigcup_{e \in E_{int}(T)} \bigcup_{x_e \in \mathrm{Crit}\, f} \mathcal{A}^{d-l}_\mathbf{Y}(x_0, x_1, \ldots, x_{i-1}, x_e, x_{i+l+1}, \ldots, x_d, T^e_1) \times \mathcal{A}^{l+1}_\mathbf{Y}(x_e, x_i, \ldots, x_{i+l}, T^e_2),$$

where in the last row the respective values of i and l are defined by e being of type (i, l).

Remark 6.18 In the last row of the boundary description in Theorem 6.17, the respective space is only non-empty if

$$\mu(x_e) = \sum_{q=i}^{i+l} \mu(x_q) + 1 - l,$$

since otherwise one of the factors, and thus the cartesian product, would be the empty set for dimensional reasons.

Proof of Theorem 6.17 In the light of Theorem 6.9, it suffices to show that the admissibility condition implies for all $e \in E_{int}(T)$ and $x_e \in \mathrm{Crit}\, f$ of the right index that

$$\mathcal{B}_\mathbf{Y}((x_0, x_e), (x_1, \ldots, x_d, x_e), T, \{e\})$$
$$\overset{!}{=} \mathcal{A}^{d-l}_\mathbf{Y}(x_0, \ldots, x_{i-1}, x_e, x_{i+l+1}, \ldots, x_d, T^e_1) \times \mathcal{A}^{l+1}_\mathbf{Y}(x_e, x_i, \ldots, x_{i+l}, T^e_2).$$

As previously discussed, this is our original motivation for the notion of admissible perturbation data. Since (6.13) and (6.14) hold, we only need to show that

$$\left(\gamma_0, \left(l_f, \gamma_f \right)_{f \in E_{int}(T_1^e)}, \gamma_1, \ldots, \gamma_{i-1}, \gamma_+, \gamma_{i+l+1}, \ldots, \gamma_d \right)$$
$$\in \mathcal{M}_{\mathbf{Y}}^d(x_0, x_1, \ldots, x_{i-1}, x_e, x_{i+l+1}, \ldots, x_d, T_1^e),$$
$$\left(\gamma_-, \left(l_f, \gamma_f \right)_{f \in E_{int}(T_2^e)}, \gamma_i, \ldots, \gamma_{i+l} \right) \in \mathcal{M}_{\mathbf{Y}}^d(x_e, x_i, \ldots, x_{i+l}, T_2^e).$$

This condition is equivalent to the following componentwise description:

$$\gamma_0 \in W^- \left(x_0, Y_{10} \left((l_f)_{f \in E_{int}(T) \setminus \{e\}} \right) \right),$$
$$(l_f, \gamma_f) \in \mathcal{M} \left(Y_{if} \left((l_g)_{g \in E_{int}(T_i^e) \setminus f} \right) \right) \qquad \forall f \in E_{int}(T_i^e),\ i \in \{1, 2\},$$
$$\gamma_j \in W^+ \left(x_j, Y_{1j} \left((l_f)_{f \in E_{int}(T_1^e)} \right) \right) \qquad \forall j \in \{1, 2, \ldots, i-1\},$$
$$\gamma_j \in W^+ \left(x_j, Y_{2(j-i+1)} \left((l_f)_{f \in E_{int}(T_2^e)} \right) \right) \qquad \forall j \in \{i, i+1, \ldots, i+l\}$$
$$\gamma_j \in W^+ \left(x_j, Y_{1(j-l)} \left((l_f)_{f \in E_{int}(T_1^e)} \right) \right) \qquad \forall j \in \{i+l+1, i+l+2, \ldots, d\},$$
$$\gamma_+ \in W^+ \left(x_e, Y_{1i} \left((l_f)_{f \in E_{int}(T_1^e)} \right) \right), \qquad \gamma_- \in W^- \left(x_e, Y_{20} \left((l_f)_{f \in E_{int}(T_2^e)} \right) \right),$$

where we put

$$\mathbf{Y}_{T_1^e} =: \left(Y_{10}, \left(Y_{1f} \right)_{f \in E_{int}(T_1^e)}, Y_{11}, \ldots, Y_{1(d-l)} \right),$$
$$\mathbf{Y}_{T_2^e} =: \left(Y_{20}, \left(Y_{2f} \right)_{f \in E_{int}(T_2^e)}, Y_{21}, \ldots, Y_{2(l+1)} \right).$$

But these conditions are a direct consequence of the admissibility of the perturbation datum $\mathbf{Y}$. They are obtained by inserting the definition of the background perturbations $\mathbf{X}^{\mathbf{Y}}$ into conditions (6.15)–(6.18). □

6.6   Constructing Universal One-Dimensional Moduli Spaces

Our next aim is to consider the union of all moduli spaces of perturbed Morse ribbon trees modelled on all possible choices of ribbon trees at once. We will construct a certain quotient space of the union of their compactifications which will finally enable us to prove that the higher order multiplications constructed in this section satisfy the defining equations of an A_∞-algebra.

The starting point is the following lemma from graph theory which can be proven by elementary methods. We omit the details.

Lemma 6.19 Let $d \geq 2$ and $T \in \text{RTree}_d$ with $k(T) = d - 3$. Then there are precisely two binary trees $T_1, T_2 \in \text{RTree}_d$ as well as unique edges $e_i \in E_{int}(T_i)$ for $i \in \{1, 2\}$ such that

$$T_1/e_1 = T = T_2/e_2.$$

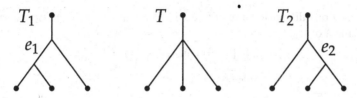

Fig. 6.5 An example of the situation of Lemma 6.19 for $d = 3$

Sketch of proof For $d = 3$, the claim is obvious, see Fig. 6.5. For arbitrary $d \geq 2$, one checks that a tree $T \in \text{RTree}_d$ with $k(T) = d - 3$ contains a unique 4-valent internal vertex while every other internal vertex is trivalent. This 4-valent vertex v can be "resolved" in precisely two different ways.

If we add another edge e_{new} to the tree whose incoming vertex is identified with v, then there are precisely two ways to connect the other edges connected to v with e_{new}. Let e_0 be the unique edge with $v_{out}(e_0) = v$ and let $e_1, e_2, e_3 \in E(T)$ denote the edges with $v_{in}(e_i) = v$ for $i \in \{1, 2, 3\}$, ordered according to the ordering induced by the given one of the leaves of T. The two different ribbon trees including e_{new} are obtained as follows:

- either we define a tree by putting

$$v_{in}(e_1) = v_{in}(e_2) = v_{out}(e_{new}) \quad \text{and} \quad v_{in}(e_3) = v_{in}(e_{new}) = v \,,$$

- or we construct a tree by putting

$$v_{in}(e_1) = v_{in}(e_{new}) = v \quad \text{and} \quad v_{in}(e_2) = v_{in}(e_3) = v_{out}(e_{new}) \,.$$

The two resulting trees are obviously different from each other and one checks that these two ways are indeed the only ways of constructing binary trees which reduce to T after collapsing a single edge. $\qquad \square$

See Fig. 6.5 for an illustration of Lemma 6.19. Theorem 6.9 and Lemma 6.19 together yield:

Corollary 6.20 *Let* $\mathbf{Y}$ *be a regular and universal perturbation datum. Let* $d \geq 2$ *and* $T \in \text{RTree}_d$ *with* $k(T) = d - 3$. *Then there are precisely two binary trees* $T_1, T_2 \in \text{RTree}_d$ *with*

$$\mathcal{A}_{\mathbf{Y}}^d(x_0, x_1, \ldots, x_d, T) \subset \partial \overline{\mathcal{A}}_{\mathbf{Y}}^d(x_0, x_1, \ldots, x_d, T_i) \qquad \forall i \in \{1, 2\} \,.$$

Let $\text{BinTree}_d \subset \text{RTree}_d$ denote the set of all binary trees with d leaves. For our choice of $\mathbf{Y}$, $x_0, x_1, \ldots, x_d \in \text{Crit } f$ and $T \in \text{BinTree}_d$, the space $\overline{\mathcal{A}}_{\mathbf{Y}}^d(x_0, x_1, \ldots, x_d, T)$ is a compact one-dimensional manifold with boundary, hence the space

$$\bigsqcup_{T \in \text{BinTree}_d} \overline{\mathcal{A}}_{\mathbf{Y}}^d(x_0, x_1, \ldots, x_d, T)$$

is a compact one-dimensional manifold with boundary as well. By Corollary 6.20, every element of

$$\bigsqcup_{\substack{T \in \mathrm{RTree}_d \\ k(T)=d-3}} \mathcal{A}_{\mathbf{Y}}^d(x_0, x_1, \ldots, x_d, T)$$

appears precisely twice as a boundary element of $\bigsqcup_{T \in \mathrm{BinTree}_d} \overline{\mathcal{A}}_{\mathbf{Y}}^d(x_0, x_1, \ldots, x_d, T)$.

Note that any compact one-dimensional manifold with boundary is diffeomorphic to a disjoint union of finitely many compact intervals and circles, so we can easily glue components of one-dimensional manifolds along their boundaries. We use this gluing procedure to define a single moduli space including elements of $\mathcal{A}_{\mathbf{Y}}^d(x_0, x_1, \ldots, x_d, T)$ for different choices of $T \in \mathrm{RTree}_d$.

Definition 6.21 Let $x_0, x_1, \ldots, x_d \in \mathrm{Crit}\, f$ with

$$\mu(x_0) = \sum_{i=1}^{d} \mu(x_i) + 3 - d$$

and let $\mathbf{Y}$ be a regular and universal perturbation datum. We introduce an equivalence relation on

$$\bigsqcup_{T \in \mathrm{BinTree}_d} \overline{\mathcal{A}}_{\mathbf{Y}}^d(x_0, x_1, \ldots, x_d, T)$$

by identifying both copies of the respective elements of $\bigsqcup_{k(T)=d-3} \mathcal{A}_{\mathbf{Y}}^d(x_0, x_1, \ldots, x_d, T)$, which exist by Corollary 6.20. With respect to this equivalence relation, we define

$$\overline{\mathcal{A}}_{\mathbf{Y}}^d(x_0, x_1, \ldots, x_d) := \bigsqcup_{T \in \mathrm{BinTree}_d} \overline{\mathcal{A}}_{\mathbf{Y}}^d(x_0, x_1, \ldots, x_d, T) \Big/ \sim ,$$

Moreover, we put

$$\mathcal{A}_{\mathbf{Y}}^d(x_0, x_1, \ldots, x_d) := \bigsqcup_{T \in \mathrm{RTree}_d} \mathcal{A}_{\mathbf{Y}}^d(x_0, x_1, \ldots, x_d, T) .$$

It is clear that the space $\overline{\mathcal{A}}_{\mathbf{Y}}^d(x_0, x_1, \ldots, x_d)$ can again be given the structure of a compact one-dimensional manifold with boundary, and that

$$\mathcal{A}_{\mathbf{Y}}^d(x_0, x_1, \ldots, x_d) \subset \mathrm{int}\left(\overline{\mathcal{A}}_{\mathbf{Y}}^d(x_0, x_1, \ldots, x_d)\right) .$$

Furthermore, if we additionally choose $\mathbf{Y}$ to be consistent, its boundary has a useful description as we will see in the next theorem.

Theorem 6.22 *Let $d \geq 2$, $x_0, x_1, \ldots, x_d \in \text{Crit } f$ with*

$$\mu(x_0) = \sum_{i=1}^{d} \mu(x_i) + 3 - d ,$$

and let **Y** *be an admissible perturbation datum. The space* $\overline{\mathcal{A}}_{\mathbf{Y}}^{d}(x_0, x_1, \ldots, x_d)$ *can be equipped with the structure of a compact one-dimensional manifold with boundary, whose boundary is given by*

$$\partial \overline{\mathcal{A}}_{\mathbf{Y}}^{d}(x_0, x_1, \ldots, x_d)$$
$$= \bigcup_{i=1}^{d} \bigcup_{l=0}^{d-i} \bigcup_{y} \mathcal{A}_{\mathbf{Y}}^{d-l}(x_0, x_1, \ldots, x_{i-1}, y, x_{i+l+1}, \ldots, x_d) \times \mathcal{A}_{\mathbf{Y}}^{l+1}(y, x_i, \ldots, x_{i+l}) ,$$

where we put $\mathcal{A}_{\mathbf{Y}}^{1}(x, y) := \widehat{\mathcal{M}}(x, y)$ *for every* $x, y \in \text{Crit } f$ *and where the union over* y *is taken over all* $y \in \text{Crit } f$ *satisfying*

$$\mu(y) = \sum_{q=i}^{i+l} \mu(x_q) + 1 - l .$$

We need another graph-theoretic lemma to prove Theorem 6.22.

Lemma 6.23 *For $d \geq 2$ let* $\widetilde{\text{RTree}}_d := \{(T, e) \mid T \in \text{RTree}_d, \ e \in E_{int}(T)\}$. *Then the following map is a bijection:*

$$G : \widetilde{\text{RTree}}_d \to \bigcup_{l=1}^{d-2} (\text{RTree}_{d-l} \times \text{RTree}_{l+1} \times \{1, 2, \ldots, d - l\}) ,$$
$$(T, e) \mapsto (T_1^e, T_2^e, i) , \quad \text{if } e \text{ is of type } (i, l).$$

Proof The injectivity of G is obvious, so we will only show its surjectivity.

Let $l \in \{1, 2, \ldots, d - 2\}$ and $(T_1, T_2, i) \in \text{RTree}_{d-l} \times \text{RTree}_{l+1} \times \{1, 2, \ldots, d - l\}$. With respect to these choices, let $T \in \text{RTree}_d$ the unique ribbon tree with

$$V(T) = (V(T_1) \setminus \{v_i(T_1)\}) \cup (V(T_2) \setminus \{v_0(T_2)\}) ,$$
$$E(T) = (E(T_1) \setminus \{e_i(T_1)\}) \cup (E(T_2) \setminus \{e_0(T_2)\}) \cup \{e_{\text{new}}\} ,$$

where e_{new} satisfies $v_{\text{in}}(e_{\text{new}}) = v_{\text{in}}(e_i(T_1))$ and $v_{\text{out}}(e_{\text{new}}) = v_{\text{out}}(e_0(T_2))$. Moreover, let $V(T)$ be ordered as follows:

$$(v_1(T_1), \ldots, v_{i-1}(T_1), v_1(T_2), \ldots, v_{l+1}(T_2), v_{i+1}(T_1), \ldots, v_{d-l}(T_1)) .$$

Then it is clear from its definition that e_{new} is of type (i, l) in T, and that

$$G(T, e_{\text{new}}) = (T_1, T_2, i) \,. \qquad\qquad \square$$

Proof of Theorem 6.22   We have discussed the manifold property before, so it only remains to describe the boundary of $\overline{\mathcal{A}}^d_{\mathbf{Y}}(x_0, x_1, \ldots, x_d)$.

From the definition of $\overline{\mathcal{A}}^d_{\mathbf{Y}}(x_0, x_1, \ldots, x_d)$ as a quotient of $\bigsqcup_{T \in \mathrm{RTree}_d} \mathcal{A}^d_{\mathbf{Y}}(x_0, x_1, \ldots, x_d, T)$, we derive

$$\partial \overline{\mathcal{A}}^d_{\mathbf{Y}}(x_0, x_1, \ldots, x_d) \subset \bigsqcup_{T \in \mathrm{RTree}_d} \partial \overline{\mathcal{A}}^d_{\mathbf{Y}}(x_0, x_1, \ldots, x_d, T) \,.$$

The boundary spaces $\partial \overline{\mathcal{A}}^d_{\mathbf{Y}}(x_0, x_1, \ldots, x_d, T)$ were described in Theorem 6.17. Note that the boundary parts of the form $\mathcal{A}^d_{\mathbf{Y}}(x_0, x_1, \ldots, x_d, T/e)$ are identified in the gluing procedure used to define $\overline{\mathcal{A}}^d_{\mathbf{Y}}(x_0, x_1, \ldots, x_d)$. All other boundary curves of any of the spaces $\overline{\mathcal{A}}^d_{\mathbf{Y}}(x_0, x_1, \ldots, x_d, T)$ are also boundary curves of $\overline{\mathcal{A}}^d_{\mathbf{Y}}(x_0, x_1, \ldots, x_d)$, so we obtain

$$\partial \overline{\mathcal{A}}^d_{\mathbf{Y}}(x_0, x_1, \ldots, x_d) = \bigcup_{T \in \mathrm{RTree}_d} \bigcup_{\substack{y_0 \in \mathrm{Crit} f \\ \mu(y_0) = \mu(x_0) - 1}} \widehat{\mathcal{M}}(x_0, y_0) \times \mathcal{A}^d_{\mathbf{Y}}(y_0, x_1, \ldots, x_d, T)$$

$$\cup \bigcup_{T \in \mathrm{RTree}_d} \bigcup_{e \in E_{int}(T)} \bigcup_{x_e \in \mathrm{Crit} f} \mathcal{A}^{d-l}_{\mathbf{Y}}(x_0, x_1, \ldots, x_{i-1}, x_e, x_{i+l+1}, \ldots, x_d, T^e_1)$$

$$\times \mathcal{A}^{l+1}_{\mathbf{Y}}(x_e, x_i, \ldots, x_{i+l}, T^e_2)$$

$$\cup \bigcup_{T \in \mathrm{RTree}_d} \bigcup_{i=1}^{d} \bigcup_{\substack{y_i \in \mathrm{Crit} f \\ \mu(y_i) = \mu(x_i) + 1}} \mathcal{A}^d_{\mathbf{Y}}(x_0, x_1, \ldots, x_{i-1}, y_i, x_{i+1}, \ldots, x_d, T) \times \widehat{\mathcal{M}}(y_i, x_i)$$

$$= \bigcup_{\substack{y_0 \in \mathrm{Crit} f \\ \mu(y_0) = \mu(x_0) - 1}} \mathcal{A}^1_{\mathbf{Y}}(x_0, y_0) \times \left(\bigcup_{T \in \mathrm{RTree}_d} \mathcal{A}^d_{\mathbf{Y}}(y_0, x_1, \ldots, x_d, T) \right)$$

$$\cup \bigcup_{T \in \mathrm{RTree}_d} \bigcup_{e \in E_{int}(T)} \bigcup_{x_e \in \mathrm{Crit} f} \mathcal{A}^{d-l}_{\mathbf{Y}}(x_0, x_1, \ldots, x_{i-1}, x_e, x_{i+l+1}, \ldots, x_d, T^e_1)$$

$$\times \mathcal{A}^{l+1}_{\mathbf{Y}}(x_e, x_i, \ldots, x_{i+l}, T^e_2)$$

$$\cup \bigcup_{i=1}^{d} \bigcup_{\substack{y_i \in \mathrm{Crit} f \\ \mu(y_i) = \mu(x_i) + 1}} \left(\bigcup_{T \in \mathrm{RTree}_d} \mathcal{A}^d_{\mathbf{Y}}(x_0, x_1, \ldots, x_{i-1}, y_i, x_{i+1}, \ldots, x_d, T) \right) \times \mathcal{A}^1_{\mathbf{Y}}(y_i, x_i)$$

$$= \bigcup_{\substack{y_0 \in \mathrm{Crit} f \\ \mu(y_0) = \mu(x_0) - 1}} \mathcal{A}^1_{\mathbf{Y}}(x_0, y_0) \times \mathcal{A}^d_{\mathbf{Y}}(y_0, x_1, \ldots, x_d)$$

$$\cup \bigcup_{T \in \mathrm{RTree}_d} \bigcup_{e \in E_{int}(T)} \bigcup_{x_e \in \mathrm{Crit} f} \mathcal{A}^{d-l}_{\mathbf{Y}}(x_0, x_1, \ldots, x_{i-1}, x_e, x_{i+l+1}, \ldots, x_d, T^e_1)$$

$$\times \mathcal{A}^{l+1}_{\mathbf{Y}}(x_e, x_i, \ldots, x_{i+l}, T^e_2)$$

$$\cup \bigcup_{\substack{i=1 \\ }}^{d} \bigcup_{\substack{y_i \in \mathrm{Crit}\, f \\ \mu(y_i)=\mu(x_i)+1}} \mathcal{A}_{\mathbf{Y}}^{d}(x_0, x_1, \ldots, x_{i-1}, y_i, x_{i+1}, \ldots, x_d) \times \mathcal{A}_{\mathbf{Y}}^{1}(y_i, x_i) \, .$$

where the unions over x_e are taken over all $x_e \in \mathrm{Crit}\, f$ satisfying

$$\mu(x_e) = \sum_{q=i}^{i+l} \mu(x_q) + 1 - l \, .$$

By Lemma 6.23, we can identify

$$\bigcup_{T \in \mathrm{RTree}_d} \bigcup_{e \in E_{int}(T)} \bigcup_{x_e \in \mathrm{Crit}\, f} \mathcal{A}_{\mathbf{Y}}^{d-l}(x_0, x_1, \ldots, x_{i-1}, x_e, x_{i+l+1}, \ldots, x_d, T_1^e)$$

$$\times \mathcal{A}_{\mathbf{Y}}^{l+1}(x_e, x_i, \ldots, x_{i+l}, T_2^e)$$

$$= \bigcup_{l=1}^{d-2} \bigcup_{(T_1,T_2) \in \mathrm{RTree}_{d-l} \times \mathrm{RTree}_{l+1}} \bigcup_{i=1}^{d-l} \bigcup_{y \in \mathrm{Crit}\, f} \mathcal{A}_{\mathbf{Y}}^{d-l}(x_0, x_1, \ldots, x_{i-1}, y, x_{i+l+1}, \ldots, x_d, T_1)$$

$$\times \mathcal{A}_{\mathbf{Y}}^{l+1}(y, x_i, \ldots, x_{i+l}, T_2)$$

$$= \bigcup_{l=1}^{d-2} \bigcup_{i=1}^{d-1} \bigcup_{y \in \mathrm{Crit}\, f} \left(\left(\bigcup_{T_1 \in \mathrm{RTree}_{d-l}} \mathcal{A}_{\mathbf{Y}}^{d-l}(x_0, x_1, \ldots, x_{i-1}, y, x_{i+l+1}, \ldots, x_d, T_1) \right) \right.$$

$$\left. \times \left(\bigcup_{T_2 \in \mathrm{RTree}_{l+1}} \mathcal{A}_{\mathbf{Y}}^{l+1}(y, x_i, \ldots, x_{i+l}, T_2^e) \right) \right)$$

$$= \bigcup_{l=1}^{d-2} \bigcup_{i=1}^{d-1} \bigcup_{y \in \mathrm{Crit}\, f} \mathcal{A}_{\mathbf{Y}}^{d-l}(x_0, x_1, \ldots, x_{i-1}, y, x_{i+l+1}, \ldots, x_d) \times \mathcal{A}_{\mathbf{Y}}^{l+1}(y, x_i, \ldots, x_{i+l}) \, ,$$

where the unions over y are taken over all $y \in \mathrm{Crit}\, f$ satisfying

$$\mu(y) = \sum_{q=i}^{i+l} \mu(x_q) + 1 - l \, . \tag{6.22}$$

Inserting this into our description of $\partial \overline{\mathcal{A}}_{\mathbf{Y}}^{d}(x_0, x_1, \ldots, x_d)$ yields:

$$\partial \overline{\mathcal{A}}_{\mathbf{Y}}^{d}(x_0, x_1, \ldots, x_d) = \bigcup_{\substack{y_0 \in \mathrm{Crit}\, f \\ \mu(y_0)=\mu(x_0)-1}} \mathcal{A}_{\mathbf{Y}}^{1}(x_0, y_0) \times \mathcal{A}_{\mathbf{Y}}^{d}(y_0, x_1, \ldots, x_d)$$

$$\cup \bigcup_{l=1}^{d-2} \bigcup_{i=1}^{d-l} \bigcup_{y \in \mathrm{Crit}\, f} \mathcal{A}_{\mathbf{Y}_1}^{d-l}(x_0, x_1, \ldots, x_{i-1}, y, x_{i+l+1}, \ldots, x_d) \times \mathcal{A}_{\mathbf{Y}_2}^{l+1}(y, x_i, \ldots, x_{i+l})$$

$$\cup \bigcup_{i=1}^{d} \bigcup_{\substack{y_i \in \mathrm{Crit}\, f \\ \mu(y_i)=\mu(x_i)+1}} \mathcal{A}_{\mathbf{Y}}^{d}(x_0, x_1, \ldots, x_{i-1}, y_i, x_{i+1}, \ldots, x_d) \times \mathcal{A}_{\mathbf{Y}}^{1}(y_i, x_i)$$

$$= \bigcup_{i=1}^{d} \bigcup_{l=0}^{d-i} \bigcup_{y \in \mathrm{Crit}\, f} \mathcal{A}_{\mathbf{Y}_1}^{d-l}(x_0, x_1, \ldots, x_{i-1}, y, x_{i+l+1}, \ldots, x_d) \times \mathcal{A}_{\mathbf{Y}_2}^{l+1}(y, x_i, \ldots, x_{i+l}),$$

where the unions over y are again taken over all $y \in \mathrm{Crit}\, f$ satisfying (6.22). $\square$

6.7 A_∞-algebra Properties of the Higher Order Multiplications

We want to conclude this chapter by discussing the A_∞-algebra properties of the maps $\mu_{d,\mathbf{X}}$ from Sect. 6.3 making use of the results from Sects. 6.5 and 6.6.

Before continuing our geometric line of argument, we present a short algebraic interlude in which we introduce the notion of an A_∞-algebra.

Our sign conventions coincide with those used in [Sei08, GJ90].

Definition 6.24 Let R be a commutative ring with unit. A graded R-bimodule $A = \bigoplus_{j \in \mathbb{Z}} A_j$ equipped with a family of homomorphisms of graded R-modules

$$\mu_n : A^{\otimes n} \to A, \qquad \deg \mu_n = 2 - n, \quad \text{for every } n \in \mathbb{N},$$

will be called an A_∞-algebra over R if the following equation is satisfied for all $r \in \mathbb{N}$ and $a_1, \ldots, a_r \in A$:

$$\sum_{\substack{n_1, n_2 \in \mathbb{N} \\ n_1 + n_2 = r+1}} \sum_{i=1}^{r+1-n_1} (-1)^{\maltese_1^{i-1}} \mu_{n_2}(a_1, \ldots, a_{i-1}, \mu_{n_1}(a_i, \ldots, a_{i+n_1-1}), a_{i+n_1}, \ldots, a_r) = 0,$$

where[1]

$$\maltese_i^j := \maltese_i^j(a_1, \ldots, a_r) := \sum_{q=i}^{j} \mu(x_q) - (j - i + 1) \tag{6.23}$$

for all $i, j \in \{1, 2, \ldots, r\}$ with $i \le j$ and where for every $k \in \mathbb{Z}$ we write $\mu(a) = k$ iff $a \in A_k$. For a fixed $r \in \mathbb{N}$ we refer to the above equation as *the rth defining equation* of the A_∞-algebra $(A, (\mu_n)_{n \in \mathbb{N}})$.

Example 6.25   1. Every differential graded algebra (DGA) over R, i.e. an associative graded algebra over R which comes with a differential satisfying the graded Leibniz rule, can be given the structure of an A_∞-algebra. We simply put μ_1 to be the differential, μ_2 to be the algebra multiplication and μ_n to be zero for

---

[1] The use of the Maltese cross for the coefficients defining the signs in this book happens in accordance with the notation of the works of Mohammed Abouzaid and Paul Seidel, e.g. [Sei08] or [Abo10]. In particular, the author distances himself from any political meaning or implication of this symbol.

every $n \geq 3$. In this case, the second defining equation reduces to the graded Leibniz rule, the third defining equation is equivalent to the associativity of the multiplication and all defining equations of higher order simply vanish on both sides.

2. James Stasheff introduced the notion of an A_∞-algebra in [Sta63a, Sta63b] and showed that the singular chain complex of a based loop space can always be equipped with the structure of an A_∞-algebra. The map μ_1 is given (up to inverting the grading) by the differential of the complex. The map μ_2 is given by the Pontryagin product, i.e. the map induced on singular chains by the composition of loops. While it is easy to see that the Pontryagin product is associative up to a chain homotopy, Stasheff introduced A_∞-algebras to show that this statement can be strongly refined.

Remark 6.26 It is possible to give a slightly more concise and elegant definition of an A_∞-algebra and all other notions defined in this section in terms of the usual coalgebra structure on the tensor algebra of A. This method is discussed by Ezra Getzler and John Jones in [GJ90] and by Thomas Tradler in [Tra08b, Tra08a].

We turn our attention back to the operations $\mu_{d,\mathbf{X}} : C^*(f)^{\otimes d} \to C^*(f)$. Theorem 6.22 is the decisive Theorem to derive relations on the coefficients of these operations.. We will combine it with the following basic fact from differential topology:

Every compact one-dimensional manifold with boundary has an even number of boundary points.

This follows immediately from the fact that a compact one-dimensional manifold with boundary is diffeomorphic to a finite union of compact intervals and circles. See [GP74, Appendix 2] for details.

In addition to the twisted intersection numbers $a_{\mathbf{Y}}^d(x_0, x_1, \ldots, x_d)$ for $d \geq 2$ from Definition 6.3, we further define for any $x_0, x_1 \in \text{Crit } f$ and any admissible perturbation datum $\mathbf{Y}$:

$$a_{\mathbf{Y}}^1(x_0, x_1) := (-1)^{n+1} n(x_0, x_1) := (-1)^{n+1} \#_{\text{alg}} \widehat{\mathcal{M}}(x_0, x_1) \,,$$

where $n(x_0, x_1)$ is the corresponding coefficient of the Morse codifferential as explained in the introduction of this chapter. See Appendix A.1 for details on these coefficients.

Corollary 6.27 *Let $d \geq 2$, $x_0, x_1, \ldots, x_d \in \text{Crit } f$ with*

$$\mu(x_0) = \sum_{i=1}^d \mu(x_i) + 3 - d$$

and let $\mathbf{Y} = (\mathbf{Y}_T)_{T \in \bigcup_{d \geq 2} \text{RTree}_d}$ be an admissible perturbation datum. Then the following congruence modulo two is true:

$$\sum_{i=1}^{d}\sum_{l=0}^{d-i}\sum_{y\in\text{Crit } f} a_{\mathbf{Y}}^{d-l}(x_0, x_1, \ldots, x_{i-1}, y, x_{i+l+1}, \ldots, x_d) \cdot a_{\mathbf{Y}}^{l+1}(y, x_i, \ldots, x_{i+l}) \equiv 0,$$

where the sum over y is taken over all $y \in \text{Crit } f$ satisfying

$$\mu(y) = \sum_{q=i}^{i+l} \mu(x_q) + 1 - l.$$

Proof By definition of the coefficients as twisted oriented intersection numbers, the following congruence modulo two holds for all $i \in \{1, 2, \ldots, d\}, l \in \{0, 1, \ldots, d - i\}$ and $y \in \text{Crit } f$ with $\mu(y) = \sum_{q=i}^{i+l} \mu(x_q) + 1 - l$:

$$\sum_{i=1}^{d}\sum_{l=0}^{d-i}\sum_{y\in\text{Crit } f} a_{\mathbf{Y}}^{d-l}(x_0, x_1, \ldots, x_{i-1}, y, x_{i+l+1}, \ldots, x_d) \cdot a_{\mathbf{Y}}^{l+1}(y, x_i, \ldots, x_{i+l})$$

$$\equiv \sum_{i=1}^{d}\sum_{l=0}^{d-i}\sum_{y\in\text{Crit } f} \left| \mathcal{A}_{\mathbf{Y}}^{d-l}(x_0, x_1, \ldots, x_{i-1}, y, x_{i+l+1}, \ldots, x_d) \right| \cdot \left| \mathcal{A}_{\mathbf{Y}}^{l+1}(y, x_i, \ldots, x_{i+l}) \right|$$

$$\equiv \sum_{i=1}^{d}\sum_{l=0}^{d-i}\sum_{y\in\text{Crit } f} \left| \mathcal{A}_{\mathbf{Y}}^{d-l}(x_0, x_1, \ldots, x_{i-1}, y, x_{i+l+1}, \ldots, x_d) \times \mathcal{A}_{\mathbf{Y}}^{l+1}(y, x_i, \ldots, x_{i+l}) \right|.$$

But Theorem 6.22 together with the abovementioned fact implies that

$$\sum_{i=1}^{d}\sum_{l=0}^{d-i}\sum_{y\in\text{Crit } f} \left| \mathcal{A}_{\mathbf{Y}}^{d-l}(x_0, x_1, \ldots, x_{i-1}, y, x_{i+l+1}, \ldots, x_d) \times \mathcal{A}_{\mathbf{Y}}^{l+1}(y, x_i, \ldots, x_{i+l}) \right|$$

$$\equiv \left| \partial \overline{\mathcal{A}}_{\mathbf{Y}}^{d}(x_0, x_1, \ldots, x_d) \right| \equiv 0 \mod 2.$$

$\square$

We want to derive a relation of the twisted intersection numbers from the one in Corollary 6.27, which is an equality of integers and not a congruence modulo two. To do so, one needs to take a look at the orientations of the moduli spaces $\mathcal{A}_{\mathbf{Y}}^{d}(x_0, x_1, \ldots, x_d, T)$ and of the compatibility of these orientations with respect to the gluing procedure that we used to define $\overline{\mathcal{A}}_{\mathbf{Y}}^{d}(x_0, x_1, \ldots, x_d)$.

Afterwards, one needs to perform a thorough sign investigation of the sum of coefficients appearing in Corollary 6.27. Since the line of argument is very technical, the proof of Theorem 6.28 is postponed to Appendix A.5.

Theorem 6.28 *Let* $d \geq 2$, $x_0, x_1, \ldots, x_d \in \mathrm{Crit}\, f$ *with*

$$\mu(x_0) = \sum_{i=1}^{d} \mu(x_i) + 3 - d$$

all and let **Y** *be an admissible perturbation datum. Then*

$$\sum_{i=1}^{d} \sum_{l=0}^{d-i} \sum_{y \in \mathrm{Crit}\, f} (-1)^{\maltese_1^{i-1}} a_\mathbf{Y}^{d-l}(x_0, x_1, \ldots, x_{i-1}, y, x_{i+l+1}, \ldots, x_d) \cdot a_\mathbf{Y}^{l+1}(y, x_i, \ldots, x_{i+l}) = 0,$$

where $\maltese_1^{i-1} := \maltese_1^{i-1}(x_1, \ldots, x_d) = \sum_{q=1}^{i-1} \mu(q) - i + 1$ *and where the sum over* y *is taken over all* $y \in \mathrm{Crit}\, f$ *satisfying* $\mu(y) = \sum_{q=i}^{i+l} \mu(x_q) + 1 - l$.

Remember that we have defined the maps $(\mu_{d,\mathbf{Y}})_{d \geq 2}$ in Definition 6.5 in terms of the numbers $a_\mathbf{Y}^d(\cdots)$. Hence, using Theorem 6.28, we are finally able to prove that these operations fulfil the defining equations of an A_∞-algebra.

Theorem 6.29 *For every admissible perturbation datum* **Y**, *the Morse cochain complex*

$$\left(C^*(f), \left(\mu_{d,\mathbf{Y}} : C^*(f)^{\otimes d} \to C^*(f)\right)_{d \in \mathbb{N}}\right)$$

is an A_∞-*algebra over* $\mathbb{Z}$, *where* $\mu_{1,\mathbf{Y}} := (-1)^{n+1}\delta : C^*(f) \to C^*(f)$ *denotes the twisted Morse codifferential.*

Proof Note that the map $\mu_{1,\mathbf{Y}} : C^*(f) \to C^*(f)$ is explicitly defined as the $\mathbb{Z}$-linear extension of

$$(x_0 \in \mathrm{Crit}\, f) \mapsto \sum_{\substack{x_1 \in \mathrm{Crit}\, f \\ \mu(x_1) = \mu(x_0) + 1}} a_\mathbf{Y}^1(x_0, x_1) x_1 \, .$$

It suffices to show that the defining equations of an A_∞-algebra hold on generators of $C^*(f)$, i.e. that for every $d \in \mathbb{N}$ and every $x_1, \ldots, x_d \in \mathrm{Crit}\, f$ the following holds:

$$\sum_{\substack{d_1, d_2 \in \mathbb{N} \\ d_1 + d_2 = d+1}} \sum_{i=1}^{d+1-d_1} (-1)^{\maltese_1^{i-1}} \mu_{d_2,\mathbf{Y}}(x_1, \ldots, x_{i-1}, \mu_{d_1,\mathbf{Y}}(x_i, \ldots, x_{i+d_1-1}), a_{i+d_1}, \ldots, x_d)$$

$$= \sum_{l=0}^{d-1} \sum_{i=1}^{d-l} (-1)^{\maltese_1^{i-1}} \mu_{d-l,\mathbf{Y}}(x_1, \ldots, x_{i-1}, \mu_{l+1,\mathbf{Y}}(x_i, \ldots, x_{i+l}), a_{i+l+1}, \ldots, x_d) = 0 \, .$$

For all $l \in \{0, 1, \ldots, d-1\}$ and $i \in \{1, 2 \ldots, d-l\}$ we compute that

$$\mu_{d-l,\mathbf{Y}}(x_1, \ldots, x_{i-1}, \mu_{l+1,\mathbf{Y}}(x_i, \ldots, x_{i+l}), x_{i+l+1}, \ldots, x_d)$$

$$= \mu_{d-l,\mathbf{Y}}(x_1, \ldots, x_{i-1}, \sum_y a_{\mathbf{Y}}^{l+1}(y, x_i, \ldots, x_{i+l}) \cdot y, x_{i+l+1}, \ldots, x_d)$$

$$= \sum_y a_{\mathbf{Y}}^{l+1}(y, x_i, \ldots, x_{i+l}) \cdot \mu_{d-l,\mathbf{Y}}(x_1, \ldots, x_{i-1}, y, x_{i+l+1}, \ldots, x_d)$$

$$= \sum_y a_{\mathbf{Y}}^{l+1}(y, x_i, \ldots, x_{i+l}) \cdot \left(\sum_{x_0} a_{\mathbf{Y}}^{d-l}(x_0 x_1, \ldots, x_{i-1}, y, x_{i+l+1}, \ldots, x_d) \cdot x_0 \right),$$

$$(6.24)$$

where the sums over x_0 and y are taken over all $x_0, y \in \mathrm{Crit}\, f$ satisfying

$$\mu(y) = \sum_{q=i}^{i+l} \mu(x_q) + 1 - l, \quad \mu(x_0) = \sum_{q=1}^{i-1} \mu(x_q) + \mu(y) + \sum_{q=i+l+1}^{d} \mu(x_q) + 2 - d + l.$$

One checks that (6.24) is the same as the sum

$$\sum_{x_0} \sum_y a_{\mathbf{Y}}^{d-l}(x_0 x_1, \ldots, x_{i-1}, y, x_{i+l+1}, \ldots, x_d) \cdot a_{\mathbf{Y}}^{l+1}(y, x_i, \ldots, x_{i+l}) \cdot x_0,$$

where this time the sums are taken over all $x_0, y \in \mathrm{Crit}\, f$ satisfying

$$\mu(y) = \sum_{q=i}^{i+l} \mu(x_q) + 1 - l, \quad \mu(x_0) = \sum_{q=1}^{d} \mu(x_q) + 1 - d. \qquad (6.25)$$

Consequently,

$$\sum_{l=0}^{d-1} \sum_{i=1}^{d-l} (-1)^{\maltese_1^{i-1}} \mu_{d-l,\mathbf{Y}}(x_1, \ldots, x_{i-1}, \mu_{l+1,\mathbf{Y}}(x_i, \ldots, x_{i+l}), a_{i+l+1}, \ldots, x_d)$$

$$= \sum_{l=0}^{d-1} \sum_{i=1}^{d-l} \sum_{x_0} \sum_y (-1)^{\maltese_1^{i-1}} a_{\mathbf{Y}}^{d-l}(x_0, x_1, \ldots, x_{i-1}, y, x_{i+l+1}, \ldots, x_d) \cdot a_{\mathbf{Y}}^{l+1}(y, x_i, \ldots, x_{i+l}) \cdot x_0$$

$$= \sum_{x_0} \left(\sum_{i=1}^{d} \sum_{l=0}^{d-i} \sum_y (-1)^{\maltese_1^{i-1}} a_{\mathbf{Y}}^{d-l}(x_0 x_1, \ldots, x_{i-1}, y, x_{i+l+1}, \ldots, x_d) \cdot a_{\mathbf{Y}}^{l+1}(y, x_i, \ldots, x_{i+l}) \right) \cdot x_0,$$

where the sums are taken over all $x_0, y \in \mathrm{Crit}\, f$ satisfying (6.25). But by Theorem 6.28, the sum inside the brackets vanishes for every $x_0 \in \mathrm{Crit}\, f$ of the right index. Therefore, the whole sum vanishes and the claim follows. □

Remark 6.30   1. In [Abo11, Sect. 3], Abouzaid describes how to construct an equivalence of A_∞-algebras between $(C^*(f), (\mu_{d,\mathbf{X}})_{d\in\mathbb{N}})$ for an admissible perturbation datum $\mathbf{X}$ and the singular cochain complex of M, seen as a DGA with the singular codifferential and the cup product. This shows that the A_∞-structure on Morse cochains corresponds to the singular cup product and particularly from the A_∞-equation for $d = 3$ that $\mu_{2,\mathbf{x}} : C^*(f) \otimes C^*(f) \to C^*(f)$ is a chain map which induces the cup product in cohomology.

2. The constructions of A_∞-structures on Morse cochain complexes can be generalized to compact manifolds with boundary and Morse functions on them whose negative gradient is an inward-pointing vector at each boundary point. See [Abo09, Sect. 4] and [Abo11, Sect. 3] for details as well as [Sch93, Sect. 4.2.3], [Aka07, Lau11] for generalities on Morse homology for manifolds with boundary.

Chapter 7
A_∞-bimodule Structures on Morse Chain Complexes

Having established A_∞-algebra structures on Morse cochain complexes $C^*(f)$, we consider their duals, namely Morse *chain* complexes. We will use this brief final chapter to show that these possess the structures of A_∞-bimodules over the A_∞-algebra $C^*(f)$.

This statement will be shown by applying a general algebraic construction for dual modules of A_∞-algebras that we present in detail in Sect. 7.1 before returning to the geometric constructions from the previous chapters in Sect. 7.2.

The notion of A_∞-bimodules naturally occurs in the discussion of Hochschild (co)homology of A_∞-algebras with coefficients, see [Mes16b]. Here, generalizing the Hochschild homology of associative algebras with coefficients in bimodules, one defines Hochschild (co)homology of an A_∞-algebra A with coefficients in an arbitrary A_∞-bimodule over A.

7.1 A_∞-bimodules over A_∞-algebras

We begin by defining the notion of an A_∞-bimodule over an A_∞-algebra and providing some examples of A_∞-algebra bimodules.

Throughout this section, let $(A, (\mu_d)_{d \in \mathbb{N}})$ be an A_∞-algebra over a commutative ring with unit R and let $\maltese_i^j$ be defined as in Sect. 6.7. We further let $\mu(x)$ denote the degree of x and put

$$\|x\| := \mu(x) - 1$$

for all $x \in A$.

The following algebraic notion was introduced by Ezra Getzler and John D.S. Jones in [GJ90] and by Martin Markl in [Mar92].

© Springer International Publishing AG, part of Springer Nature 2018
S. Mescher, *Perturbed Gradient Flow Trees and A_∞-algebra Structures in Morse Cohomology*, Atlantis Studies in Dynamical Systems 6,
https://doi.org/10.1007/978-3-319-76584-6_7

Definition 7.1 A graded R-bimodule M equipped with a family of maps $(\mu_{r,s}^M)_{r,s\in\mathbb{N}_0}$ where

$$\mu_{r,s}^M : A^{\otimes r} \otimes M \otimes A^{\otimes s} \to M$$

is an R-bimodule homomorphism with $\deg \mu_{r,s}^M = 1 - r - s$ for all $r, s \in \mathbb{N}_0$ will be called *an A_∞-bimodule over A* if the following equation holds for all $r, s \in \mathbb{N}_0$, $m \in M$ and $a_1, \dots, a_{r+s} \in A$, where $\mu_M(m)$ denotes the degree of m:

$$\sum_{\substack{r_1\in\mathbb{N}_0,\, r_2\in\mathbb{N} \\ r_1+r_2=r+1}} \sum_{i=1}^{r_1} (-1)^{\maltese_1^{i-1}} \mu_{r_1,s}^M(a_1, \dots, a_{i-1}, \mu_{r_2}(a_i, \dots, a_{i+r_2-1}), a_{i+r_2}, \dots, a_r, m, a_{r+1}, \dots, a_{r+s})$$

$$+ \sum_{\substack{r_1,r_2\in\mathbb{N}_0\, s_1,s_2\in\mathbb{N}_0 \\ r_1+r_2=r\; s_1+s_2=s}} (-1)^{\maltese_1^{r_1}} \mu_{r_1,s_1}^M(a_1, \dots, a_{r_1}, \mu_{r_2,s_2}^M(a_{r_1+1}, \dots, a_r, m, a_{r+1}, \dots, a_{r+s_2}), \dots, a_{r+s})$$

$$+ \sum_{\substack{s_1\in\mathbb{N}_0,\, s_2\in\mathbb{N} \\ s_1+s_2=s+1}} \sum_{j=1}^{s_1} (-1)^{\maltese_1^{r-j+1}+\mu_M(m)}$$

$$\mu_{r,s_1}^M(a_1, \dots, a_r, m, a_{r+1}, \dots, a_{r+j-1}, \mu_{s_2}(a_{r+j}, \dots, a_{r+j+s_2-1}), a_{r+j+s_2}, \dots, a_{r+1}) = 0 \,.$$

We refer to this equation as the *defining equation of type (r, s) for the A_∞-bimodule M*.

Note that the defining equation of type $(0, 0)$ is equivalent to the map

$$\mu^M := \mu_{0,0}^M : M \to M$$

being a differential of degree $+1$ on M.

Remark 7.2 One checks that the graded module $A[1]$ is an A_∞-bimodule over A, where $(A[1])_j = A_{j+1}$, with the operations defined by

$$\mu_{r,s}^{A[1]} : A^{\otimes r} \otimes A[1] \otimes A^{\otimes s} \to A[1] \,,$$

$$\mu_{r,s}^{A[1]}(a_1, \dots, a_r, a_0, a_{r+1}, \dots, a_{r+s}) = \mu_{r+s+1}(a_1, \dots, a_r, a_0, a_{r+1}, \dots, a_{r+s}) \,,$$

for all $r, s \in \mathbb{N}_0$.

Next we construct a slightly more sophisticated example which is taken from [Abo10, Sect. 4], where no complete proof is given. We show that $A \otimes A$ will admit the structure of an A_∞-bimodule over A if we equip it with the grading

$$\mu_{A\otimes A}(b_1 \otimes b_2) = \|b_1\| + \|b_2\| \,,$$

i.e. if we actually consider the product grading on $A[1] \otimes A[1]$. Define maps $\mu_{r,s}^\otimes$ by

$$\mu_{r,s}^\otimes : A^{\otimes r} \otimes (A \otimes A) \otimes A^{\otimes s} \to A \otimes A \, , \quad \mu_{r,s}^\otimes = 0 \quad \text{for } r,s > 0 \, ,$$

$$\mu_{r,0}^\otimes : A^{\otimes r} \otimes (A \otimes A) \to A \otimes A \, ,$$

$$a_1 \otimes \cdots \otimes a_r \otimes (b_1 \otimes b_2) \mapsto \mu_{r+1}(a_1, \ldots, a_r, b_1) \otimes b_2 \quad \text{for } r > 0 \, ,$$

$$\mu_{0,s}^\otimes : (A \otimes A) \otimes A^{\otimes s} \to A \otimes A \, ,$$

$$(b_1 \otimes b_2) \otimes a_1 \otimes \cdots \otimes a_s \mapsto (-1)^{\|b_1\|} b_1 \otimes \mu_{s+1}(b_2, a_1, \ldots, a_s) \quad \text{for } s > 0 \, ,$$

$$\mu_{0,0}^\otimes : A \otimes A \to A \otimes A \, , \quad b_1 \otimes b_2 \mapsto \mu_1(b_1) \otimes b_2 + (-1)^{\|b_1\|} b_1 \otimes \mu_1(b_2) \, .$$

Theorem 7.3 ([Abo10], Proposition 4.7) $(A \otimes A, (\mu_{r,s}^\otimes)_{r,s \in \mathbb{N}_0})$ is an A_∞-bimodule over A.

Proof By definition, $\mu_{r,s}^\otimes$ vanishes if both $r > 0$ and $s > 0$ and so does the left-hand side of the defining equation of an A_∞-bimodule. If $r = 0$, $s = 0$, the defining equation is equivalent to $\mu_{0,0}^\otimes \circ \mu_{0,0}^\otimes = 0$, which is clear from the definition of $\mu_{0,0}^\otimes$ as a product differential.

Assume that $r > 0$ and $s = 0$. Then the third sum on the left-hand side of the defining equation of type $(r, 0)$ vanishes by definition, while for every $a_1, \ldots, a_r, b_1, b_2 \in A$ the first sum is given by:

$$\sum_{\substack{r_1 \in \mathbb{N}_0, r_2 \in \mathbb{N} \\ r_1 + r_2 = r+1}} \sum_{i=1}^{r_1} (-1)^{\maltese_1^{i-1}} \mu_{r_1,0}^\otimes(a_1, \ldots, a_{i-1}, \mu_{r_2}(a_i, \ldots, a_{i+r_2-1}), a_{i+r_2}, \ldots, a_r, b_1 \otimes b_2)$$

$$= \sum_{\substack{r_1 \in \mathbb{N}_0, r_2 \in \mathbb{N} \\ r_1 + r_2 = r+1}} \sum_{i=1}^{r_1} (-1)^{\maltese_1^{i-1}} \mu_{r_1+1}(a_1, \ldots, a_{i-1}, \mu_{r_2}(a_i, \ldots, a_{i+r_2-1}), a_{i+r_2}, \ldots, a_r, b_1) \otimes b_2$$

$$= \sum_{\substack{r_1, r_2 \in \mathbb{N} \\ r_1 + r_2 = r+2}} \sum_{i=1}^{r_1-1} (-1)^{\maltese_1^{i-1}} \mu_{r_1}(a_1, \ldots, a_{i-1}, \mu_{r_2}(a_i, \ldots, a_{i+r_2-1}), \ldots, a_r, b_1) \otimes b_2 \, . \quad (7.1)$$

The reader may notice the similarity of this sum to the left-hand side of the $(r + 1)$-st defining equation of the A_∞-algebra A for the elements $a_1, \ldots, a_r, b_1 \in A$ tensorized with b_2, except that some parts of the defining equation are still missing. We will see that these parts are contained in the second sum of the defining equation of type (r, s) which we next write down explicitly.

$$\sum_{\substack{r_1, r_2 \in \mathbb{N}_0 \\ r_1 + r_2 = r}} (-1)^{\maltese_1^{r_1}} \mu_{r_1,0}^\otimes(a_1, \ldots, a_{r_1}, \mu_{r_2,0}^\otimes(a_{r_1+1}, \ldots, a_r, b_1 \otimes b_2))$$

$$= (-1)^{\maltese_1^r} \mu_{r,0}^\otimes(a_1, \ldots, a_r, \mu_{0,0}^\otimes(b_1 \otimes b_2)) + \mu_{0,0}^\otimes(\mu_{r,0}^\otimes(a_1, \ldots, a_r, b_1 \otimes b_2))$$

$$+ \sum_{\substack{r_1, r_2 \in \mathbb{N} \\ r_1 + r_2 = r}} (-1)^{\maltese_1^{r_1}} \mu_{r_1,0}^\otimes(a_1, \ldots, a_{r_1}, \mu_{r_2,0}^\otimes(a_{r_1+1}, \ldots, a_r, b_1 \otimes b_2))$$

$$= (-1)^{\maltese_1^r} \mu_{r,0}^\otimes(a_1, \ldots, a_r, \mu_1(b_1) \otimes b_2 + (-1)^{\|b_1\|} b_1 \otimes \mu_1(b_2))$$

$$+ \mu_{0,0}^\otimes(\mu_{r+1}(a_1, \ldots, a_r, b_1) \otimes b_2)$$

$$+ \sum_{\substack{r_1,r_2\in\mathbb{N}\\r_1+r_2=r}} (-1)^{\maltese_1^{r_1}} \mu_{r_1+1}(a_1,\dots,a_{r_1},\mu_{r_2+1}(a_{r_1+1},\dots,a_r,b_1))\otimes b_2$$

$$= (-1)^{\maltese_1^r}\mu_{r+1}(a_1,\dots,a_r,\mu_1(b_1))\otimes b_2 + (-1)^{\maltese_1^r+\|b_1\|}\mu_{r+1}(a_1,\dots,a_r,b_1)\otimes\mu_1(b_2)$$

$$+ \mu_1(\mu_{r+1}(a_1,\dots,a_r,b_1))\otimes b_2 + (-1)^{\|\mu_{r+1}(a_1,\dots,a_r,b_1)\|}\mu_{r+1}(a_1,\dots,a_r,b_1)\otimes\mu_1(b_2)$$

$$+ \sum_{\substack{r_1,r_2\geq 2\\r_1+r_2=r+2}} (-1)^{\maltese_1^{r_1-1}}\mu_{r_1}(a_1,\dots,a_{r_1-1},\mu_{r_2}(a_{r_1},\dots,a_r,b_1))\otimes b_2$$

$$= \sum_{r_1\geq 2,r_2\in\mathbb{N}} (-1)^{\maltese_1^{r_1-1}}\mu_{r_1}(a_1,\dots,a_{r_1-1},\mu_{r_2}(a_{r_1},\dots,a_r,b_1))\otimes b_2 \tag{7.2}$$

$$+ \mu_1(\mu_{r+1}(a_1,\dots,a_r,b_1))\otimes b_2 ,$$

where we have used the fact that

$$\|\mu_{r+1}(a_1,\dots,a_r,b_1)\| = \sum_{q=1}^{r}\|a_q\| + \|b_1\| + 1 = \maltese_1^r + \|b_1\| + 1 ,$$

since $\deg\mu_{r+1} = 1-r$. Thus, we have shown that for $A\otimes A$, the left-hand side of the defining equation for A_∞-bimodules is given by the sum of (7.1) and (7.2), which amounts to:

$$\sum_{\substack{r_1,r_2\in\mathbb{N}\\r_1+r_2=r+2}} \sum_{i=1}^{r_1-1}(-1)^{\maltese_1^{i-1}}\mu_{r_1}(a_1,\dots,a_{i-1},\mu_{r_2}(a_i,\dots,a_{i+r_2-1}),a_{i+r_2},\dots,a_r,b_1)\otimes b_2$$

$$+ \sum_{\substack{r_1,r_2\in\mathbb{N}\\r_1+r_2=r+2}} (-1)^{\maltese_1^{r_1-1}}\mu_{r_1}(a_1,\dots,a_{r_1-1},\mu_{r_2}(a_{r_1},\dots,a_r,b_1))\otimes b_2$$

$$= \sum_{\substack{r_1,r_2\in\mathbb{N}\\r_1+r_2=r+2}} \sum_{i=1}^{r_1}(-1)^{\maltese_1^{i-1}}\mu_{r_1}(a_1,\dots,a_{i-1},\mu_{r_2}(a_i,\dots,a_{i+r_2-1}),a_{i+r_2},\dots,a_r,b_1)\otimes b_2 .$$

The latter sum equals the left-hand side of the A_∞-equation for $r+1$ tensorized with b_2. Since A is an A_∞-algebra, the sum vanishes.

We omit the remaining case, $r = 0$ and $s > 0$, since it is proven analogously. $\square$

The following theorem shows that the dual R-bimodule of an A_∞-bimodule over A can always be equipped with the structure of an A_∞-bimodule over the same A_∞-algebra. We will apply this dual bimodule structure in Sect. 7.2 to Morse chain and cochain complexes.

Theorem 7.4 ([Tra08b], Lemma 3.9) *Let* $\left(M, (\mu_{r,s}^M)_{r,s\in\mathbb{N}_0}\right)$ *be an A_∞-bimodule over* R *and let* $M^* = \mathrm{Hom}_R(M, R)$ *denote its dual R-bimodule. Then* $\left(M^{-*}, (\mu_{r,s}^*)_{r,s\in\mathbb{N}_0}\right)$ *is an A_∞-bimodule over A, where M^{-*} denotes M^* with inverted grading, i.e.*

$$M^{-j} := (M^{-*})^j := \operatorname{Hom}_R(M_{-j}, R) \, ,$$

and where

$$\mu_{r,s}^* : A^{\otimes r} \otimes M^{-*} \otimes A^{\otimes s} \to M^{-*}$$

is for all $r, s \in \mathbb{N}_0$, $a_1, \ldots, a_{r+s} \in A$, $m \in M$ *and* $m^* \in M^{-*}$ *defined by*

$$\left(\mu_{r,s}^*(a_1, \ldots, a_r, m^*, a_{r+1}, \ldots, a_{r+s})\right)(m) = (-1)^{\ddagger_{r,s}} m^* \left(\mu_{s,r}^M(a_{r+1}, \ldots, a_{r+s}, m, a_1, \ldots, a_r)\right)$$

with

$$\ddagger_{r,s} := \ddagger_{r,s}(a_1, \ldots, a_r, m^*, a_{r+1}, \ldots, a_{r+s}, m)$$
$$:= \maltese_1^r \cdot \left(\maltese_{r+1}^{r+s} + \mu_{M^*}(m^*) + \mu_M(m)\right) + \mu_{M^*}(m^*) + 1 \, .$$

Proof (We repeat Tradler's proof and add a sign computation since we use different sign conventions than Tradler.)

We exhibit the operations $\left(\mu_{r,s}^*\right)_{r,s \in \mathbb{N}_0}$ to fulfill the A_∞-bimodule equation. We need to show for all $r, s \in \mathbb{N}_0$, $a_1, \ldots, a_{r+s} \in A$, $m \in M$ and $m^* \in M^{-*}$ that

$$\sum_{r_1+r_2=r+1} \sum_{i=1}^{r_1} (-1)^{\maltese_1^{i-1}} \mu_{r_1,s}^*(a_1, \ldots, \mu_{r_2}(a_i, \ldots, a_{i+r_2-1}), a_{i+r_2}, \ldots, a_r, m^*, a_{r+1}, \ldots, a_{r+s})(m)$$

$$+ \sum_{r_1+r_2=r} \sum_{s_1+s_2=s} (-1)^{\maltese_1^{r_1}} \mu_{r_1,s_1}^*(a_1, \ldots, \mu_{r_2,s_2}^*(a_{r_1+1}, \ldots, a_r, m^*, a_{r+1}, \ldots, a_{r+s_2}), \ldots, a_{r+s})(m)$$

$$+ \sum_{s_1+s_2=s+1} \sum_{j=1}^{s_1} (-1)^{\maltese_1^{r+j-1}+\mu(m^*)}$$

$$\mu_{r,s_1}^*(a_1, \ldots, a_r, m^*, a_{r+1}, \ldots, a_{r+j-1}, \mu_{s_2}(a_{r+j}, \ldots, a_{r+j+s_2-1}), a_{r+j+s_2}, \ldots, a_{r+1})(m) \overset{!}{=} 0 \, .$$

where $\mu(m^*) := \mu_{M^*}(m^*)$. By definition of the $\mu_{r,s}^*$, the left-hand side of this equation is given by

$$\sum_{r_1+r_2=r+1} \sum_{i=1}^{r_1} (-1)^{\maltese_1^{i-1}+\ddagger_{r_1,s}(a_1,\ldots,\mu_{r_2}(a_i,\ldots,a_{i+r_2-1}),\ldots,a_r,m^*,a_{r+1},\ldots,a_{r+s},m)}$$

$$m^* \left(\mu_{s,r_1}^M(a_{r+1}, \ldots, a_{r+s}, m, a_1, \ldots, \mu_{r_2}(a_i, \ldots, a_{i+r_2-1}), \ldots, a_r)\right)$$

$$+ \sum_{r_1+r_2=r} \sum_{s_1+s_2=s} (-1)^{\maltese_1^{r_1}+\ddagger_{r_1,s_1}(a_1,\ldots,a_{r_1},\mu_{r_2,s_2}^*(a_{r_1+1},\ldots,a_r,m^*,a_{r+1},\ldots,a_{r+s_2}),a_{r+s_2+1},\ldots,a_{r+s},m)}$$

$$\left(\mu_{r_2,s_2}^*(a_{r_1+1}, \ldots, a_r, m^*, a_{r+1}, \ldots, a_{r+s_2})\right) \left(\mu_{s,r_1}^M(a_{r+s_2+1}, \ldots, a_{r+s}, m, a_1, \ldots, a_{r_1})\right)$$

$$+ \sum_{s_1+s_2=s+1} \sum_{j=1}^{s_1} (-1)^{\maltese_1^{r+j-1}+\mu(m^*)+\ddagger_{r,s_1}(a_1,\ldots,a_r,m^*,a_{r+1},\ldots,\mu_{s_2}(a_{r+j},\ldots,a_{r+j+s_2-1}),\ldots,a_{r+s},m)}$$

$$m^* \left(\mu_{s_1,r}^M(a_{r+1}, \ldots, a_{r+j-1}, \mu_{s_2}(a_{r+j}, \ldots, a_{r+j+s_2-1}), a_{r+j+s_2}, \ldots, a_{r+s}, m, a_1, \ldots, a_r)\right)$$

$$= \sum_{r_1+r_2=r+1} \sum_{i=1}^{r_1} (-1)^{\maltese_1^{i-1}+\maltese_{r_1,s}(a_1,\ldots,\mu_{r_2}(a_i,\ldots,a_{i+r_2-1}),\ldots,a_r,m^*,a_{r+1},\ldots,a_{r+s},m)}$$

$$m^*\left(\mu_{s,r_1}^M(a_{r+1},\ldots,a_{r+s},m,a_1,\ldots,\mu_{r_2}(a_i,\ldots,a_{i+r_2-1}),\ldots,a_r)\right)$$

$$+ \sum_{r_1+r_2=r} \sum_{s_1+s_2=s}$$

$$(-1)^{\maltese_1^{r_1}+\maltese_{r_1,s_1}(a_1,\ldots,\mu_{r_2,s_2}^*(a_{r_1+1},\ldots,m^*,\ldots,a_{r+s_2}),\ldots,a_{r+s},m)+\maltese_{r_2,s_2}(a_{r_1+1},\ldots,m^*,\ldots,a_{r+s_2},\mu_{s_1,r_1}^M(a_{r+s_2+1},\ldots,m,\ldots,a_{r_1}))}$$

$$m^*\left(\mu_{s_2,r_2}^M(a_{r+1},\ldots,a_{r+s_2},\mu_{s_1,r_1}^M(a_{r+s_2+1},\ldots,a_{r+s},m,a_1,\ldots,a_{r_1}),a_{r_1+1},\ldots,a_r)\right)$$

$$+ \sum_{s_1+s_2=s+1} \sum_{j=1}^{s_1}(-1)^{\maltese_1^{r+j-1}+\mu(m^*)+\maltese_{r,s_1}(a_1,\ldots,a_r,m^*,a_{r+1},\ldots,\mu_{s_2}(a_{r+j},\ldots,a_{r+j+s_2-1}),\ldots,a_{r+s},m)}$$

$$m^*\left(\mu_{s_1,r}^M(a_{r+1},\ldots,a_{r+j-1},\mu_{s_2}(a_{r+j},\ldots,a_{r+j+s_2-1}),a_{r+j+s_2},\ldots,a_{r+s},m,a_1,\ldots,a_r)\right) \,,$$

Clearly this sum vanishes for every $m^* \in M^{-*}$ if and only if

$$\sum_{r_1+r_2=r+1} \sum_{i=1}^{r_1}(-1)^{S_1}\mu_{s,r_1}^M(a_{r+1},\ldots,a_{r+s},m,a_1,\ldots,\mu_{r_2}(a_i,\ldots,a_{i+r_2-1}),\ldots,a_r)$$

$$+ \sum_{r_1+r_2=r} \sum_{s_1+s_2=s}(-1)^{S_2}\mu_{s_2,r_2}^M(a_{r+1},\ldots,\mu_{s_1,r_1}^M(a_{r+s_2+1},\ldots,a_{r+s},m,a_1,\ldots,a_{r_1}),\ldots,a_r)$$

$$+ \sum_{s_1+s_2=s+1} \sum_{j=1}^{s_1}(-1)^{S_3}\mu_{s_1,r}^M(a_{r+1},\ldots,\mu_{s_2}(a_{r+j},\ldots,a_{r+j+s_2-1}),\ldots,a_{r+s},m,a_1,\ldots,a_r) = 0 \,,$$

where we denote the exponents of (-1) in the previous computation by S_1, S_2 and S_3, respectively. Since $(M, (\mu_{r,s}^M)_{r,s\in\mathbb{N}_0})$ is an A_∞-bimodule over A, the latter equality holds true if we can show that

$$\begin{aligned}
S_1 &\equiv \maltese_{r+1}^{r+s} + \maltese_1^{i-1} + \mu_M(m) + k \,, \\
S_2 &\equiv \maltese_{r+1}^{r+s_2} + k \,, \qquad S_3 \equiv \maltese_{r+1}^{r+j-1} + k \,,
\end{aligned} \tag{7.3}$$

for some $k \in \mathbb{Z}$, where '$\equiv$' denotes congruence modulo two. In this case, multiplying the desired equation with $(-1)^k$ yields the defining equation of type (r, s) for the A_∞-bimodule M. So it only remains to check the signs.

Concerning S_1, we compute that

$$\ddagger_{r_1,s}(a_1,\ldots,\mu_{r_1}(a_i,\ldots,a_{i+r_1-1}),\ldots,a_r,m^*,a_{r+1},\ldots,a_{r+s},m)$$

$$=\left(\sum_{q=1}^{i-1}\|a_q\|+\|\mu_{r_2}(a_i,\ldots,a_{i+r_2-1})\|+\sum_{q=i+r_2}^{r}\|a_q\|\right)\left(\maltese_{r+1}^{r+s}+\mu(m^*)+\mu_M(m)\right)$$

$$+\mu(m^*)+1$$

$$\equiv\left(\maltese_1^r+1\right)\left(\maltese_{r+1}^{r+s}+\mu(m^*)+\mu_M(m)\right)+\mu(m^*)+1$$

$$\equiv\maltese_{r+1}^{r+s}+\mu_M(m)+\mu(m^*)+\ddagger_{r,s}(a_1,\ldots,a_r,m^*,a_{r+1},\ldots,a_{r+s},m)$$

and therefore by definition of S_1:

$$S_1\equiv\maltese_{r+1}^{r+s}+\maltese_1^{i-1}+\mu_M(m)+k_0$$

if we put $k_0:=\mu(m^*)+\ddagger_{r,s}$. This shows the first line of (7.3). Considering the sign given by S_2, we first compute

$$\ddagger_{r_1,s_1}(a_1,\ldots,a_{r_1},\mu_{r_2,s_2}^*(a_{r_1+1},\ldots,a_r,m^*,a_{r+1},\ldots,a_{r+s_2}),a_{r+s_2+1},\ldots,a_{r+s},m)$$

$$\equiv\maltese_1^{r_1}\left(\maltese_{r+s_2+1}^{r+s}+\mu\left(\mu_{r_2,s_2}^*(a_{r_1+1},\ldots,a_r,m^*,a_{r+1},\ldots,a_{r+s_2})\right)+\mu_M(m)\right)$$

$$+\mu\left(\mu_{r_2,s_2}^*(a_{r_1+1},\ldots,a_r,m^*,a_{r+1},\ldots,a_{r+s_2})\right)+1$$

$$\equiv\maltese_1^{r_1}\left(\maltese_{r+s_2+1}^{r+s}+\maltese_{r_1+1}^{r+s_2}+\mu(m^*)+\mu_M(m)+1\right)+\maltese_{r_1+1}^{r+s_2}+\mu(m^*)$$

$$\equiv\maltese_1^{r_1}\left(\maltese_{r_1+1}^{r+s}+\mu(m^*)+\mu_M(m)+1\right)+\maltese_{r_1+1}^{r+s_2}+\mu(m^*)$$

$$\equiv\maltese_1^{r+s_2}+\maltese_1^{r_1}\left(\maltese_{r_1+1}^{r+s}+\mu(m^*)+\mu_M(m)\right)+\mu(m^*).$$

Furthermore, notice that

$$\ddagger_{r_2,s_2}(a_{r_1+1},\ldots,a_r,m^*,a_{r+1},\ldots,a_{r+s_2},\mu_{s_1,r_1}^M(a_{r+s_2+1},\ldots,a_{r+s},m,a_1,\ldots,a_{r_1}))$$

$$\equiv\maltese_{r_1+1}^{r}\left(\maltese_{r+1}^{r+s_2}+\mu(m^*)+\mu_M\left(\mu_{s_1,r_1}^M(a_{r+s_2+1},\ldots,a_{r+s},m,a_1,\ldots,a_{r_1})\right)\right)+\mu(m^*)+1$$

$$\equiv\maltese_{r_1+1}^{r}\left(\maltese_{r+1}^{r+s}+\maltese_1^{r_1}+\mu(m^*)+\mu_M(m)+1\right)+\mu(m^*)+1$$

$$\equiv\maltese_{r_1+1}^{r}+\maltese_1^{r_1}\maltese_{r_1+1}^{r}+\maltese_{r_1+1}^{r}\left(\maltese_{r+1}^{r+s}+\mu(m^*)+\mu_M(m)\right)+\mu(m^*)+1.$$

Combining these last two computations, the definition of S_2 yields

$$S_2\equiv\maltese_1^{r_1}+\maltese_1^{r+s_2}+\maltese_{r_1+1}^{r}+\maltese_1^{r_1}\left(\maltese_{r_1+1}^{r+s}+\mu(m^*)+\mu_M(m)\right)+\maltese_1^{r_1}\maltese_{r_1+1}^{r}$$

$$+\maltese_{r_1+1}^{r}\left(\maltese_{r+1}^{r+s}+\mu(m^*)+\mu_M(m)\right)+1$$

$$\equiv\maltese_{r+1}^{r+s_2}+\maltese_1^{r}\left(\maltese_{r+1}^{r+s}+\mu(m^*)+\mu_M(m)\right)+1\equiv\maltese_{r+1}^{r+s_2}+k_0,$$

implying the first congruence from the second line of (7.3). Finally, for S_3 we consider:

$\ddagger_{r,s_1}(a_1,\ldots,a_r,m^*,a_{r+1},\ldots,\mu_{s_2}(a_{r+j},\ldots,a_{r+j+s_2-1}),\ldots,a_{r+s},m)$

$\equiv \maltese_1^r\left(\maltese_{r+1}^{r+j-1} + \|\mu_{s_2}(a_{r+j},\ldots,a_{r+j+s_2-1})\| + \maltese_{r+j+s_2}^{r+s} + \mu(m^*) + \mu_M(m)\right) + \mu(m^*) + 1$

$\equiv \maltese_1^r\left(\maltese_{r+1}^{r+s} + 1 + \mu(m^*) + \mu_M(m)\right) + \mu(m^*) + 1 \equiv \maltese_1^r + \mu(m^*) + k_0 \, .$

We consequently obtain that $S_3 \equiv \maltese_{r+1}^{r+j-1} + k_0$. Thus, we have shown all three congruences of (7.3) with $k = k_0$ and therefore completed the proof.   □

7.2 Dualization of the Higher Order Multiplications

Throughout this section, we choose and fix an admissible perturbation datum $\mathbf{X}$ *and consider* $(C^*(f), (\mu_d)_{d\in\mathbb{N}})$ *as an* A_∞-*algebra over* $\mathbb{Z}$, *where* $\mu_d := \mu_{d,\mathbf{X}}$ *for every* $d \in \mathbb{N}$. *We further denote* $\|x\| = \mu(x) - 1$ *for every* $x \in \mathrm{Crit}\, f$ *and let* '$\equiv$' *always denote congruence modulo two.*

We want to apply the algebraic construction of dual A_∞-bimodules from the previous section to the case

$$A = C^*(f) \quad \text{and} \quad M = C^*(f)[1]\,,$$

where, see Remark 7.2,

$${}'(C^*(f)[1])_k = C^{k+1}(f) \quad \forall k \in \mathbb{Z}\,.$$

Since M is compact and critical points of Morse functions are isolated, f has finitely many critical points. Thus, $C^*(f)$ has finite rank and there is a canonical isomorphism

$$\left(C^*(f)\right)^* = \mathrm{Hom}_\mathbb{Z}(C^*(f); \mathbb{Z}) \cong C_*(f)\,. \tag{7.4}$$

We let $C_{-*}(f)[1]$ denote $C_*(f)$ with inverted grading and shifted degree, i.e.

$$(C_{-*}(f)[1])_k = C_{-(k+1)}(f) \quad \forall k \in \mathbb{Z}\,.$$

By Theorem 7.4, we can equip $C_{-*}(f)[1] \cong (C^*(f)[1])^{-*}$ with the structure of an A_∞-bimodule over $C^*(f)$ by purely algebraic means. In the present case, the abstractly given dual bimodule structure on $C_{-*}(f)[1]$ is given in a much more explicit way in terms of the A_∞-algebra multiplications $(\mu_d)_{d\in\mathbb{N}}$.

Let $x \in \mathrm{Crit}\, f$. Depending on whether we view it as an element of $C_*(f)$ or $C^*(f)$, we will write:

$$x \in C^*(f) \quad \text{and} \quad x^* \in C_*(f)\,.$$

This notation might seem awkward at first sight, but it will help us in keeping the dualization viewpoint in mind.

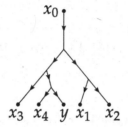

Fig. 7.1 An illustration of the perturbed Morse ribbon trees defining $\mu_{2,2}^*$

Theorem 7.5 $\left(C_{-*}(f)[1], \left(\mu_{r,s}^*\right)_{r,s\in\mathbb{N}_0}\right)$ *is an A_∞-bimodule over $C^*(f)$, where for all $r, s \in \mathbb{N}_0$ the map*

$$\mu_{r,s}^* : C^*(f)^{\otimes r} \otimes C_{-*}(f)[1] \otimes C^*(f)^{\otimes s} \to C_{-*}(f)[1]$$

is defined as the $\mathbb{Z}$-linear extension of

$$\mu_{r,s}^*\left(x_1, \ldots, x_r, y^*, x_{r+1}, \ldots, x_{r+s}\right) = (-1)^{\mu(y)} \sum_z a_{\mathbf{X}}^{r+s+1}(y, x_{r+1}, \ldots, x_{r+s}, z, x_1, \ldots, x_r) \cdot z^*$$

for $x_1, \ldots, x_{r+s}, y \in \mathrm{Crit}\, f$, where the sum is taken over all $z \in \mathrm{Crit}\, f$ satisfying

$$- \mu(z) = -\mu(y) + \sum_{q=1}^{r+s} \mu(x_q) + 1 - r - s . \tag{7.5}$$

See Fig. 7.1 for an illustration of the coefficients of the $\mu_{r,s}^*$.

Proof It suffices to show that the operations $\left(\mu_{r,s}^*\right)_{r,s\in\mathbb{N}_0}$ coincide with the dual bimodule operations from Theorem 7.4 on every product of critical points of f. By definition of the dual operations, viewing $C_{-*}(f)[1]$ as $(C^*(f)[1])^{-*}$, the following holds for all $x_1, \ldots, x_{r+s}, y, z \in \mathrm{Crit}\, f$:

$$\left(\mu_{r,s}^*\left(x_1, \ldots, x_r, y^*, x_{r+1}, \ldots, x_{r+s}\right)\right)(z) = (-1)^{\ddagger_{r,s}} y^*\left(\mu_{r,s}(x_1, \ldots, x_r, z, x_{r+1}, \ldots, x_{r+s})\right)$$

$$= (-1)^{\ddagger_{r,s}} y^*\left(\sum_{y_0} a_{\mathbf{X}}^{r+s+1}(y_0, x_{r+1}, \ldots, x_{r+s}, z, x_1, \ldots, x_r) \cdot y_0\right)$$

$$= (-1)^{\ddagger_{r,s}} \sum_{y_0} a_{\mathbf{X}}^{r+s+1}(y_0, x_{r+1}, \ldots, x_{r+s}, z, x_1, \ldots, x_r) \cdot y^*(y_0)$$

$$= (-1)^{\ddagger_{r,s}} a_{\mathbf{X}}^{r+s+1}(y, x_{r+1}, \ldots, x_{r+s}, z, x_1, \ldots, x_r) ,$$

where the sums over y_0 are taken over all $y_0 \in \mathrm{Crit}\, f$ with

$$\mu(y_0) = \sum_{q=1}^{r+s} \mu(x_q) + \mu(z) + 1 - r - s$$

and where

$$\ddagger_{r,s} = \maltese_1^r(\maltese_{r+1}^{r+s} + \mu_{C_*(f)[1]}(y^*) + \mu_{C^*(f)[1]}(z)) + \mu_{C_*(f)[1]}(y^*) + 1\,.$$

It follows from (7.5) that

$$\mu(y) \equiv \mu(z) + \sum_{q=1}^{r+s} \mu(x_q) - 1 - r - s \;\;\Leftrightarrow\;\; \|y\| \equiv \|z\| + \maltese_1^{r+s} + 1\,.$$

Since $\mu_{C^*(f)[1]}(z) = \|z\|$ and $\mu_{C_*(f)[1]}(y^*) = \|y\|$, we derive that

$$\begin{aligned}
\ddagger_{r,s} &\equiv \maltese_1^r\left(\maltese_{r+1}^{r+s} + \|y\| + \|z\|\right) + \|y\| + 1 \\
&\equiv \maltese_1^r\left(\maltese_{r+1}^{r+s} + \maltese_1^{r+s} + 2\|z\| + 1\right) + \mu(y) \\
&\equiv \maltese_1^r\left(\maltese_1^r + 1\right) + \mu(y) \equiv \mu(y)\,.
\end{aligned}$$

Since this computation holds for every $z \in \mathrm{Crit}\, f$, it follows that:

$$\mu_{r,s}^*\,(x_1,\dots,x_r,y,x_{r+1},\dots,x_{r+s}) = (-1)^{\mu(y)} \sum_z a_{\mathbf{X}}^{r+s+1}(y,x_{r+1},\dots,x_{r+s},z,x_1,\dots,x_r)\cdot z^*\,,$$

where the sum is taken over all $z \in \mathrm{Crit}\, f$ satisfying (7.5). $\square$

Remark 7.6 The bimodule operations $\left(\mu_{r,s}^*\right)_{r,s\in\mathbb{N}_0}$ are an A_∞-refinement of the chain-level cap product in Morse homology. One derives from the defining equation of an A_∞-bimodule for $(0, 1)$ that the operation

$$\mu_{0,1}^* : C_{-*}(f) \otimes C^*(f) \to C_{-*}(f)$$

is a chain map. Although we do note prove it here, the induced map on (co)homology level is well-known to correspond to the homology cap product under the identification of Morse and singular homology.

We want to investigate what happens if we iterate the dual bimodule structures for Morse chain and cochain complexes. While $C^*(f)$ and $(C_{-*}(f))^{-*}$ are obviously isomorphic as free abelian groups, it is a priori not clear whether the dual A_∞-bimodule structure on $(C_{-*}(f)[1])^{-*}$ that is induced by the dual bimodule structure on $C_{-*}(f)$ coincides with the one that $C^*(f)[1]$ possesses as an A_∞-bimodule over $C^*(f)$ that we discussed in Remark 7.2.

By Theorem 7.4, the A_∞-bimodule operations on $(C_{-*}(f)[1])^{-*}$ are induced by the ones on $C_{-*}(f)[1]$, i.e. by the family $(\mu_{r,s}^*)_{r,s\in\mathbb{N}_0}$, while the canonical A_∞-bimodule structure on $C^*(f)[1]$ is given by the higher order multiplications, i.e. by $\mu_{r,s} = \mu_{r+s+1}$ in shorthand notation, see the remarks after Definition 7.1.

In the following, we will always identify $C^*(f)[1] \cong (C_{-*}(f)[1])^{-*}$ as free abelian groups without further mentioning.

Definition 7.7 We call the A_∞-bimodule structure on $C^*(f)[1]$ which, by Theorem 7.4, is induced by the family $\left(\mu_{r,s}^*\right)_{r,s\in\mathbb{N}_0}$ *the bidual* A_∞-*bimodule structure* and denote it by

$$\left(\mu_{r,s}^{**}: C^*(f)^{\otimes r}\otimes C^*(f)[1]\otimes C^*(f)^{\otimes s}\to C^*(f)[1]\right)_{r,s\in\mathbb{N}_0}.$$

Proposition 7.8 *The bidual* A_∞-*bimodule structure on* $C^*(f;\mathbb{Z}_2)[1]$ *coincides with the one given by the* A_∞-*algebra multiplications up to a sign twist. More precisely, it holds that*

$$\mu_{r,s}^{**}(x_1,\ldots,x_r,x_0^{**},x_{r+1},\ldots,x_{r+s}) = (-1)^{\maltese_1^{r+s}+1}\mu_{r+s+1}(x_1,\ldots,x_r,x_0,x_{r+1},\ldots,x_{r+s})$$

for all $r,s\in\mathbb{N}_0$, $x_0,x_1,\ldots,x_{r+s}\in\mathrm{Crit}\,f$, *where again*

$$\maltese_1^{r+s} = \sum_{q=1}^{r+s}\mu(x_q) - r - s.$$

Proof Depending on whether we consider $x\in\mathrm{Crit}\,f$ as an element of $C^*(f)[1]$, $C_{-*}(f)[1]$ or $(C_{-*}(f)[1])^{-*}$, we write

$$x\in C^*(f)[1], \quad x^*\in C_{-*}(f)[1] \quad\text{and}\quad x^{**}\in(C_{-*}(f)[1])^{-*}.$$

In this notation it suffices to show that the following holds for all $r,s\in\mathbb{N}_0$ and every $x_0,x_1,\ldots,x_{r+s}\in\mathrm{Crit}\,f$:

$$\mu_{r,s}^{**}(x_1,\ldots,x_r,x_0^{**},x_{r+1},\ldots,x_{r+s}) \overset{!}{=} \mu_{r,s}(x_1,\ldots,x_r,x_0,x_{r+1},\ldots,x_{r+s}).$$
$$(7.6)$$

We will explicitly compute the left-hand side of Eq. (7.6). Let $y\in\mathrm{Crit}\,f$. By definition of the bidual A_∞-bimodule structure we derive:

$$\mu_{r,s}^{**}(x_1,\ldots,x_r,x_0^{**},x_{r+1},\ldots,x_{r+s})(y^*) = (-1)^{\ddagger_{r,s}}x_0^{**}(\mu_{s,r}^*(x_{r+1},\ldots,x_{r+s},y^*,x_1,\ldots,x_r)),$$

where

$$\ddagger_{r,s} = \maltese_1^r\left(\maltese_{r+1}^{r+s} + \mu_{C^*(f)[1]}(x_0^{**}) + \mu_{C_{-*}(f)[1]}(y^*)\right) + \mu_{C^*(f)[1]}(x_0^{**}) + 1.$$

In analogy with the corresponding part of the proof of Theorem 7.5, one shows that

$$\ddagger_{r,s} \equiv \mu(x_0),$$

such that

$$\mu_{r,s}^{**}(x_1, \ldots, x_r, x_0^{**}, x_{r+1}, \ldots, x_{r+s})(y^*)$$
$$= (-1)^{\mu(x_0)} x_0^{**}(\mu_{s,r}^*(x_{r+1}, \ldots, x_{r+s}, y^*, x_1, \ldots, x_r))$$
$$= (-1)^{\mu(x_0)+\mu(y)} \sum_z a_{\mathbf{X}}^{r+s+1}(y, x_1, \ldots, x_r, z, x_{r+1}, \ldots, x_{r+s}) x_0^{**}(z^*)$$
$$= (-1)^{\mu(x_0)+\mu(y)} a_{\mathbf{X}}^{r+s+1}(y, x_1, \ldots, x_r, x_0, x_{r+1}, \ldots, x_{r+s}) \,,$$

where the sums over z are taken over all $z \in \mathrm{Crit}\, f$ satisfying

$$\mu(y) = \sum_{q=1}^{r+s} \mu(x_q) + \mu(z) + 1 - r - s \,.$$

By construction, it particularly holds that

$$\mu(y) = \sum_{q=1}^{r+s} \mu(x_q) + \mu(x_0) + 1 - r - s \,,$$

which implies that

$$\mu(x_0) + \mu(y) \equiv \maltese_1^{r+s} + 1 \,.$$

To avoid the evaluation at a fixed y^*, we use this last computation and rewrite the above formula as

$$\mu_{r,s}^{**}(x_1, \ldots, x_r, x_0^{**}, x_{r+1}, \ldots, x_{r+s}) = (-1)^{\maltese_1^{r+s}+1} \sum_y a_{\mathbf{X}}^{r+s+1}(y, x_1, \ldots, x_r, x_0, x_{r+1}, \ldots, x_{r+s}) y^{**}$$
$$= (-1)^{\maltese_1^{r+s}+1} \sum_y a_{\mathbf{X}}^{r+s+1}(y, x_1, \ldots, x_r, x_0, x_{r+1}, \ldots, x_{r+s}) y$$
$$= (-1)^{\maltese_1^{r+s}+1} \mu_{r+s+1}(x_1, \ldots, x_r, x_0, x_{r+1}, \ldots, x_{r+s}) \,,$$

which is what we had to show. $\square$

Appendix A
Orientations and Sign Computations for Perturbed Morse Ribbon Trees

In the course of the main part of this book, we have mentioned the necessity of orientations on the Morse-theoretic moduli spaces involved on several occasions. We have defined the maps $\mu_{d,\mathbf{Y}}$ in terms of the twisted oriented intersection numbers of the form $a_{\mathbf{Y}}^d(x_0, x_1, \ldots, x_d)$, which requires a choice of orientations on the spaces $\mathcal{A}_{\mathbf{Y}}^d(x_0, x_1, \ldots, x_d)$.

For clarity's sake, we refrained from explaining the details in the main part of this book. Nevertheless, we will provide the constructions and explanations in this appendix. It contains several tedious, but indispensable computations related to orientations on Morse-theoretic moduli spaces and the coefficients $a_{\mathbf{Y}}^d(x_0, x_1, \ldots, x_d)$.

In Sect. A.1 we define orientations on spaces of (unperturbed) semi-infinite and finite-length Morse trajectories by using identifications of these moduli spaces with subsets of the target manifold. Afterwards, we investigate the compatibilities of the chosen orientations with Morse-theoretic gluing maps.

Section A.4 extends the constructions of the aforegoing section to spaces of perturbed Morse trajectories. The orientations on the perturbed trajectories are then used to define orientations on the moduli spaces of type $\mathcal{A}_{\mathbf{Y}}^d(x_0, x_1, \ldots, x_d, T)$, filling the gaps in the definition of the coefficients $a_{\mathbf{Y}}^d(x_0, x_1, \ldots, x_d)$. A similar discussion is given by Abouzaid in [Abo11, Sect. 8], .

Eventually, Sect. A.5 provides all missing computations for the proof of Theorem 6.28. In other words it concludes the proof of the maps $\mu_{d,\mathbf{Y}}$ fulfilling the defining equations of an A_∞-algebra by a number of long, but straightforward computations. This section is strongly oriented on [Abo09, Appendix C], which discusses the unperturbed analogue of Theorem 6.28.

In contrast to the more general approach of coherent orientations as in [Sch93, FH93], we will pursue the classic finite-dimensional approach as presented in [Sch93, Appendix B], [BH04, Sect. 7.1], [Jos08, Sect. 6.6] or [Sch05, Sect. 6.6].

Let M be an n-dimensional smooth closed oriented manifold and $f \in C^\infty(M)$ be a Morse function. Furthermore, let g be a Riemannian metric on M such that (f, g) is a Morse–Smale pair. For $x \in \mathrm{Crit} f$ we denote

© Springer International Publishing AG, part of Springer Nature 2018
S. Mescher, *Perturbed Gradient Flow Trees and A∞-algebra Structures in Morse Cohomology*, Atlantis Studies in Dynamical Systems 6, https://doi.org/10.1007/978-3-319-76584-6

$$W^u(x) := W^u(x, f) := W^u(x, f, g), \quad W^s(x) := W^s(x, f) := W^s(x, f, g),$$
$$\mathcal{W}^u(x) := \mathcal{W}^u(x, f) := \mathcal{W}^u(x, f, g), \quad \mathcal{W}^s(x) := \mathcal{W}^s(x, f) := \mathcal{W}^s(x, f, g).$$

Let always '$\equiv$' denote the congruence modulo two of two integers.

A.1 Orientations on Morse Trajectory Spaces

By the Stable/Unstable Manifold Theorem from Morse theory, see [BH04, Theorem 4.2], both the unstable and the stable manifolds of a negative gradient flow of a Morse function on a complete manifold without boundary are diffeomorphic to open balls, hence orientable.

We orient the unstable and stable manifolds with respect to $f \in C^\infty(M)$ as follows:

- We choose an arbitrary orientation on $W^u(x)$ for every $x \in \operatorname{Crit} f$.
- We equip every $W^s(x)$ with the unique orientation such that the orientation on $T_x M$ which is induced by the splitting

$$T_x M = T_x W^s(x) \oplus T_x W^u(x)$$

coincides with the given orientation on M for every $x \in \operatorname{Crit} f$. In other words, such that a positive basis of $T_x W^s(x)$ followed by a positive basis of $T_x W^u(x)$ is a positive basis of $T_x M$.

Throughout this book, we assume such orientations on the unstable and stable manifolds to be chosen and fixed.

We have mentioned in Remark 2.4 that evaluation at the endpoints of trajectories induces diffeomorphisms $\mathcal{W}^u(x) \cong W^u(x)$ and $\mathcal{W}^s(x) \cong W^s(x)$.

Throughout this book, we equip the spaces $\mathcal{W}^u(x)$ and $\mathcal{W}^s(x)$ for every $x \in \operatorname{Crit} f$ with the unique orientations which make the endpoint evaluations

$$\mathcal{W}^u(x) \overset{\cong}{\to} W^u(x) \quad and \quad \mathcal{W}^s(x) \overset{\cong}{\to} W^s(x), \quad \gamma \mapsto \gamma(0),$$

orientation-preserving.

The spaces of finite-length trajectories of the negative gradient flow of f are even easier to orient. By construction, there is a diffeomorphism

$$\mathcal{M}(f, g) \overset{\cong}{\to} [0, +\infty) \times M, \quad (l, \gamma) \mapsto (l, \gamma(0)). \tag{A.1}$$

We equip $[0, +\infty)$ with the canonical and $[0, +\infty) \times M$ with the product orientation.

Throughout this book, we equip the space $\mathcal{M}(f, g)$ with the unique orientation, such that the diffeomorphism in (A.1) is orientation-preserving if and only if n is odd. In other words, such that the sign of the diffeomorphism in (A.1) is given by $(-1)^{n+1}$.

This choice of orientation for the space of finite-length trajectories might seem awkward at first sight. For technical reasons, this choice of orientation will be the right one to establish the A_∞-equations for the Morse-theoretic operations on $C^*(f)$.

Remember from the introduction of Chap. 2 that the codifferential of the Morse cochain complex $\delta : C^*(f) \to C^*(f)$ is given as the $\mathbb{Z}$-linear extension of

$$\delta(x) = \sum_{\substack{z \in \mathrm{Crit} f \\ \mu(z) = \mu(x)+1}} n(z, x) \cdot z$$

for $x \in \mathrm{Crit} f$, where $n(z, x) := \#_{\mathrm{alg}} \widehat{\mathcal{M}}(z, x)$, i.e. the oriented intersection number of the zero-dimensional manifold $\widehat{\mathcal{M}}(z, x)$. (We will discuss oriented intersection numbers in Sect. A.4.)

The last types of Morse trajectory spaces that we use in this book are moduli spaces of parametrized and unparametrized Morse trajectories starting and ending in fixed critical points, i.e. moduli spaces of type $\mathcal{M}(x, y) := \mathcal{M}(x, y, g)$ and $\widehat{\mathcal{M}}(x, y) := \widehat{\mathcal{M}}(x, y, g)$ for $x, y \in \mathrm{Crit} f$ which we defined in the introduction of Chap. 2. We will proceed in strict analogy with [Sch93, Appendix B] and define orientations on these moduli spaces.

These orientations are important for our considerations, since, as mentioned in the introduction of Chap. 2, the codifferential of the Morse cochain complex $\delta : C^*(f) \to C^*(f)$ is given as the $\mathbb{Z}$-linear extension of

$$\delta(x) = \sum_{\substack{z \in \mathrm{Crit} f \\ \mu(z) = \mu(x)+1}} n(z, x) \cdot z$$

for $x \in \mathrm{Crit} f$, where $n(z, x) := \#_{\mathrm{alg}} \widehat{\mathcal{M}}(z, x)$, i.e. the oriented intersection number of the zero-dimensional manifold $\widehat{\mathcal{M}}(z, x)$. (We will discuss oriented intersection numbers in Sect. A.4.)

For any $x, y \in \mathrm{Crit} f$, there is an inclusion

$$\iota : \mathcal{M}(x, y) \to \mathcal{W}^u(x) \times \mathcal{W}^s(y) , \quad \gamma \mapsto \left(\gamma|_{(-\infty,0]}, \gamma|_{[0,+\infty)}\right) .$$

One checks without difficulties that ι is a smooth embedding. Let $\gamma \in \mathcal{M}(x, y)$ and put $(\gamma_1, \gamma_2) := \iota(\gamma)$. The following sequence is a short exact sequence of vector spaces:

$$0 \longrightarrow T_\gamma \mathcal{M}(x, y) \overset{i}{\longrightarrow} T_{\gamma_1} \mathcal{W}^u(x) \oplus T_{\gamma_2} \mathcal{W}^s(y) \overset{p}{\longrightarrow} T_{\gamma(0)} M \longrightarrow 0 , \quad (\mathrm{A.2})$$

where $\gamma_1 := \gamma|_{(-\infty,0]}, \gamma_2 := \gamma|_{[0,+\infty)}$,

$$\begin{aligned} i(\xi) &= \left(\xi|_{(-\infty,0]}, -\xi|_{[0,+\infty)}\right) , \\ p(\xi_1, \xi_2) &= \xi_1(0) + \xi_2(0) . \end{aligned} \quad (\mathrm{A.3})$$

The injectivity of i is clear while the surjectivity of p follows from the Morse–Smale property, i.e. because all unstable and stable manifolds intersect transversely. Given a short exact sequence of vector spaces

$$0 \to V_0 \overset{i}{\to} V_1 \overset{p}{\to} V_2 \to 0$$

and orientations on V_1 and V_2, these orientations induce a well-defined orientation on V_0 by using the following convention: A basis of V_0 is positive if and only if the image of this basis under i followed by the preimage of a positive basis of V_2 is a positive basis on V_1.

For all $x, y \in \mathrm{Crit} f$ with $\mu(x) > \mu(y)$, we equip $\mathcal{M}(x, y)$ with the orientation induced by the orientation on M, the chosen orientations on $\mathcal{W}^u(x)$ and $\mathcal{W}^s(y)$ and the short exact sequence (A.2).

Let $x, y \in \mathrm{Crit} f$ with $\mathcal{M}(x, y) \neq \varnothing$ and $x \neq y$. Since f is decreasing along the trajectories of its negative gradient flow, this implies: $f(x) > f(y)$. Let $a \in \mathbb{R}$ be a regular value of f with

$$f(x) > a > f(y).$$

Then the following map is well-defined and smooth:

$$\iota_a : \widehat{\mathcal{M}}(x, y) \to \mathcal{M}(x, y), \quad \hat{\gamma} \mapsto \gamma_a, \tag{A.4}$$

where γ_a denotes the unique element of the equivalence class $\hat{\gamma}$ satisfying

$$f(\gamma_a(0)) = a.$$

For $\hat{\gamma} \in \widehat{\mathcal{M}}(x, y)$ let $(v_1, \ldots, v_N)$ be a basis of $T_{\hat{\gamma}}\widehat{\mathcal{M}}(x, y)$. Then

$$\left(-\nabla f \circ \gamma, (D\iota_a)_{\hat{\gamma}}(v_1), \ldots, (D\iota_a)_{\hat{\gamma}}(v_N) \right)$$

is a basis of $T_{\gamma_a}\mathcal{M}(x, y)$, see [Sch93, Appendix B] or [Sch05, Sect. 2.5].

We call $(v_1, \ldots, v_N)$ a *positive* basis of $T_{\hat{\gamma}}\widehat{\mathcal{M}}(x, y)$ if and only if the thus-induced basis of $T_{\gamma_a}\mathcal{M}(x, y)$ is positive with respect to the orientation defined above.

For every $x, y \in \mathrm{Crit} f$ with $f(x) > f(y)$, this definition of positive bases yields a well-defined orientation of $\widehat{\mathcal{M}}(x, y)$. This orientation is independent of the choice of regular value a and we will always consider $\widehat{\mathcal{M}}(x, y)$ as equipped with this orientation.

We want to investigate how the Morse-theoretic gluing maps that we used in Chap. 2 behave with respect to the orientations on Morse trajectory spaces. Before we do so, we discuss the relation between Morse trajectory spaces with respect to the functions f and $-f$ that will be useful in the discussing of gluing maps and orientations.

We will derive the result for positive semi-infinite Morse trajectories from the result for negative semi-infinite Morse trajectories by identifying positive half-trajectories of the negative gradient flow of f with negative half-trajectories of its positive gradient flow.

We make the simple observation that the critical points of the Morse functions f and $-f$ coincide. In the following, we will define all negative and positive gradient flow lines with respect to the same given Riemannian metric on M.

The following map is a smooth diffeomorphism for every $x \in \mathrm{Crit} f$:

$$\varphi_s : \mathcal{W}^s(x, f) \to \mathcal{W}^u(x, -f) , \qquad (\varphi_s(\gamma))(t) := \gamma(-t) \;\; \forall t \in (-\infty, 0] .$$

We equip every $\mathcal{W}^u(x, -f)$ with the unique orientation that makes the map φ_s orientation-preserving.

Moreover, we equip every $\mathcal{W}^s(x, -f)$ with the complementary orientation, i.e. the unique orientation induced by the following splitting:

$$T_x M = T_x \mathcal{W}^s(x, -f) \oplus T_x \mathcal{W}^u(x, -f) \cong T_{\gamma_x} \mathcal{W}^s(x, -f) \oplus T_{\gamma_x} \mathcal{W}^u(x, -f) ,$$

where γ_x denotes both the constant positive and negative semi-infinite Morse trajectory, $\gamma_x(t) = t$ for every t.

In addition to the map φ_s, we consider the following smooth diffeomorphism:

$$\varphi_u : \mathcal{W}^u(x, f) \to \mathcal{W}^s(x, -f) , \qquad (\varphi_u(\gamma))(t) = \gamma(-t) \;\; \forall t \in [0, +\infty) .$$

Lemma A.1 *With respect to the chosen orientations, φ_u is orientation-preserving if and only if $(n + 1)\mu(x)$ is even.*

Proof We will write $\varphi_u(\mathcal{W}^u(x, f))$ for $\mathcal{W}^s(x, -f)$ whenever we want to consider it as equipped the orientation induced by φ_u.

By definition, φ_u is orientation-preserving if and only if the following splitting induces a positive orientation on $T_x M$:

$$T_x M \cong T_{\gamma_x} \varphi_u(\mathcal{W}^u(x, f)) \oplus T_{\gamma_x} \mathcal{W}^u(x, -f) , \qquad (A.5)$$

where γ_x again denotes the respective constant trajectory. The following map is an orientation-preserving diffeomorphism:

$$\varphi_u^{-1} \times \varphi_s^{-1} : \varphi_u(\mathcal{W}^u(x, f)) \times \mathcal{W}^u(x, -f) \to \mathcal{W}^u(x, f) \times \mathcal{W}^s(x, f) .$$

Thus, there is an orientation-preserving isomorphism

$$T_{\gamma_x} \varphi_u(\mathcal{W}^u(x, f)) \oplus T_{\gamma_x} \mathcal{W}^u(x, -f) \cong T_{\gamma_x} \mathcal{W}^u(x, f) \oplus T_{\gamma_x} \mathcal{W}^s(x, f) .$$

Moreover, there are isomorphisms

$$T_{\gamma_x} \mathcal{W}^u(x, f) \oplus T_{\gamma_x} \mathcal{W}^s(x, f) \cong T_{\gamma_x} \mathcal{W}^s(x, f) \oplus T_{\gamma_x} \mathcal{W}^u(x, f) \cong T_x M ,$$

where the latter one is by definition orientation-preserving and the former one is a permutation of factors. It is orientation-preserving if and only if the following number is even:

$$\dim \mathcal{W}^s(x, f) \cdot \dim \mathcal{W}^u(x, f) = (n - \mu(x))\mu(x) \equiv (n+1)\mu(x) \,.$$

Consequently, φ_u induces the complementary orientation in (A.5) iff $(n+1)\mu(x)$ is even. □

Another way of stating the content of Lemma A.1 is the following: If $\mathcal{W}^s(x, -f)$ is equipped with the orientation complementary to the one of $\mathcal{W}^u(x, -f)$, then the sign of the diffeomorphism $\varphi_u : \mathcal{W}^u(x, f) \to \mathcal{W}^s(x, -f)$ will be given by

$$\mathrm{sign}\varphi_u = (-1)^{(n+1)\mu(x)} \,.$$

We can also identify spaces of the form $\mathcal{M}(x, y)$ with trajectories of the positive gradient flow of f. For clarity's sake, we will denote $\mathcal{M}(x, y)$ by $\mathcal{M}(x, y, f)$ for all $x, y \in \mathrm{Crit} f$ and put

$$\mathcal{M}(y, x, -f) := \left\{ \gamma \in C^\infty(\mathbb{R}, M) \ \middle| \ \dot{\gamma} - \nabla^g f \circ \gamma = 0, \ \lim_{s \to -\infty} \gamma(s) = y, \ \lim_{s \to +\infty} \gamma(s) = x \right\} \,.$$

Since $(-f, g)$ is obviously a Morse–Smale pair if (f, g) is and since

$$\mu_{\mathrm{Morse}}(x, -f) = n - \mu(x)$$

for every $x \in \mathrm{Crit} f$, the space $\mathcal{M}(y, x, -f)$ is a smooth manifold with

$$\dim \mathcal{M}(y, x, -f) = (n - \mu(y)) - (n - \mu(x)) = \mu(x) - \mu(y) \,.$$

For every $x, y \in \mathrm{Crit} f$ the following map is a smooth diffeomorphism:

$$\psi : \mathcal{M}(x, y, f) \to \mathcal{M}(y, x, -f) \,, \qquad (\psi(\gamma))(t) := \gamma(-t) \quad \forall t \in \mathbb{R} \,.$$

Lemma A.2 *The map* $\psi : \mathcal{M}(x, y, f) \to \mathcal{M}(y, x, -f)$ *is orientation-preserving if and only if* $(\mu(x) + 1)\mu(y)$ *is even.*

Proof The rows of the following diagram are short exact sequences and one checks without difficulties that the diagram commutes:

$$
\begin{array}{ccccccccc}
0 & \longrightarrow & T_\gamma \mathcal{M}(x, y, f) & \overset{i}{\longrightarrow} & T_{\gamma_1}\mathcal{W}^u(x, f) \oplus T_{\gamma_2}\mathcal{W}^s(y, f) & \overset{p}{\longrightarrow} & T_{\gamma(0)}M & \longrightarrow & 0 \\
& & \cong \downarrow {\scriptstyle (D\psi)_\gamma} & & \cong \downarrow {\scriptstyle V} & & \cong \downarrow {\scriptstyle -\mathrm{id}_{T_{\gamma(0)}M}} & & \\
0 & \longrightarrow & T_{\psi(\gamma)}\mathcal{M}(y, x, -f) & \overset{i}{\longrightarrow} & T_{\bar{\gamma}_2}\mathcal{W}^u(y, -f) \oplus T_{\bar{\gamma}_1}\mathcal{W}^s(x, -f) & \overset{p}{\longrightarrow} & T_{\gamma(0)}M & \longrightarrow & 0,
\end{array}
$$

where $\gamma_1 := \gamma|_{(-\infty, 0]}$, $\gamma_2 := \gamma|_{[0, +\infty)}$, $\bar{\gamma}_1 := \varphi_u(\gamma_1)$ and $\bar{\gamma}_2 := \varphi_s(\gamma_2)$. Here, i and p are defined as in (A.3) and the middle vertical map V is given by

$$V(\xi_1, \xi_2) = \left(-(D\varphi_s)_{\gamma_2}(\xi_2), -(D\varphi_u)_{\gamma_1}(\xi_1)\right) \,.$$

Since the diagram commutes, it follows that

$$\mathrm{sign}\psi = \mathrm{sign}(D\psi)_{\gamma} = (\mathrm{sign}V)\left(\mathrm{sign}\left(-\mathrm{id}_{T_{\gamma(0)}M}\right)\right) = (-1)^{n} \cdot \mathrm{sign}V \,. \qquad (A.6)$$

To determine the sign of V, we note that V factorizes as

$$T_{\gamma_1}\mathcal{W}^u(x,f) \oplus T_{\gamma_2}\mathcal{W}^s(y,f) \xrightarrow{V_1} T_{\gamma_2}\mathcal{W}^s(y,f) \oplus T_{\gamma_1}\mathcal{W}^u(x,f) \xrightarrow{V_2}$$
$$T_{\bar{\gamma}_2}\mathcal{W}^u(y,-f) \oplus T_{\bar{\gamma}_1}\mathcal{W}^s(x,-f) \xrightarrow{V_3} T_{\bar{\gamma}_2}\mathcal{W}^u(y,-f) \oplus T_{\bar{\gamma}_1}\mathcal{W}^s(x,-f) \,,$$

where V_1 is the map which transposes the two factors, V_2 is given by applying $(D\varphi_s)_{\gamma_2}$ and $(D\varphi_u)_{\gamma_1}$ and $V_3 = -\mathrm{id}$. Since $V = V_3 \circ V_2 \circ V_1$, it follows that

$$\mathrm{sign}V = \mathrm{sign}V_1 \cdot \mathrm{sign}V_2 \cdot \mathrm{sign}V_3 \,.$$

The sign of V_1 is given by the parity of

$$\left(\dim T_{\gamma_1}\mathcal{W}^u(x,f)\right) \cdot \left(\dim T_{\gamma_2}\mathcal{W}^s(y,f)\right) = \mu(x)(n - \mu(y)) \,.$$

The sign of V_2 is given by

$$\mathrm{sign}V_2 = \mathrm{sign}(D\varphi_s)_{\gamma_2} \cdot \mathrm{sign}(D\varphi_u)_{\gamma_1} = \mathrm{sign}(D\varphi_u)_{\gamma_1} = (-1)^{(n+1)\mu(x)} \,,$$

where we used Lemma A.1 and the fact that φ_s is orientation-preserving. The sign of V_3 is obviously given by the parity of

$$\dim T_{\bar{\gamma}_2}\mathcal{W}^u(y,-f) + \dim T_{\bar{\gamma}_1}\mathcal{W}^s(x,-f) = n - \mu(y) + \mu(x) \,.$$

We conclude that the sign of V is given by the parity of

$$\mu(x)(n - \mu(y)) + (n+1)\mu(x) + n - \mu(y) + \mu(x) \equiv \mu(x)\mu(y) + n + \mu(y)$$
$$\equiv (\mu(x)+1)\mu(y) + n \,.$$

We derive from (A.6) that

$$\mathrm{sign}\psi = (-1)^{n}(-1)^{(\mu(x)+1)\mu(y)+n} = (-1)^{(\mu(x)+1)\mu(y)}.$$

$\square$

One observes that the map $\psi : \mathcal{M}(x,y,f) \to \mathcal{M}(y,x,-f)$ induces a well-defined map

$$\widehat{\psi} : \widehat{\mathcal{M}}(x,y,f) \to \widehat{\mathcal{M}}(y,x,-f) \,,$$

which assigns to each $\hat{\gamma}$ the equivalence class of $\psi(\gamma)$, where γ is a representative of $\hat{\gamma}$. Since ψ is a smooth diffeomorphism, one derives that $\widehat{\psi}$ is a smooth diffeomorphism as well.

Proposition A.3 *The map* $\widehat{\psi} : \widehat{\mathcal{M}}(x, y, f) \rightarrow \widehat{\mathcal{M}}(y, x, -f)$ *is orientation-preserving if and only if* $(\mu(x) + 1)\mu(y)$ *is odd.*

Proof For every regular value a of f with $f(x) > a > f(y)$, the following diagram obviously commutes, where ι_a is defined as in (A.4):

$$
\begin{array}{ccc}
\widehat{\mathcal{M}}(x, y, f) & \xrightarrow{\;\widehat{\psi}\;} & \widehat{\mathcal{M}}(y, x - f) \\
\downarrow{\iota_a} & & \downarrow{\iota_a} \\
\mathcal{M}(x, y, f) & \xrightarrow{\;\psi\;} & \mathcal{M}(y, x, -f)
\end{array}
\qquad (A.7)
$$

Let $\hat{\gamma} \in \widehat{\mathcal{M}}(x, y, f)$ and let $(v_1, \ldots, v_N)$ be a positive basis of $T_{\hat{\gamma}}\widehat{\mathcal{M}}(x, y, f)$. We need to determine the sign of the basis

$$
\left((D\widehat{\psi})_{\hat{\gamma}}(v_1), \ldots, (D\widehat{\psi})_{\hat{\gamma}}(v_N) \right)
$$

of $T_{\widehat{\psi}(\hat{\gamma})}\widehat{\mathcal{M}}(y, x, -f)$. By definition of the orientation on $\widehat{\mathcal{M}}(y, x, -f)$, this basis is positive if and only if the following is a positive basis of $T_{(\iota_a \circ \widehat{\psi})(\hat{\gamma})}\mathcal{M}(y, x, -f)$:

$$
\left(\nabla f \circ \left(\iota_a \circ \widehat{\psi} \right)(\hat{\gamma}), \left(D \left(\iota_a \circ \widehat{\psi} \right) \right)_{\hat{\gamma}}(v_1), \ldots, \left(D \left(\iota_a \circ \widehat{\psi} \right) \right)_{\hat{\gamma}}(v_N) \right) \qquad (A.8)
$$
$$
= \left(\nabla f \circ \psi(\gamma_a), \left(D \left(\psi \circ \iota_a \right) \right)_{\hat{\gamma}}(v_1), \ldots, \left(D \left(\psi \circ \iota_a \right) \right)_{\hat{\gamma}}(v_N) \right),
$$

where the equality follows from the commutativity of diagram (A.7).

Since ψ inverts the flow direction, the following obviously holds for every $\gamma \in \mathcal{M}(x, y, f)$:

$$
(D\psi)_\gamma(-\nabla f \circ \gamma) = \nabla f \circ \psi(\gamma).
$$

Hence the basis in (A.8) is given by

$$
\left((D\psi)_{\gamma_a}(-\nabla f \circ \gamma_a), \left(D \left(\psi \circ \iota_a \right) \right)_{\hat{\gamma}}(v_1), \ldots, \left(D \left(\psi \circ \iota_a \right) \right)_{\hat{\gamma}}(v_N) \right)
$$
$$
= \left((D\psi)_{\gamma_a}(-\nabla f \circ \gamma_a), (D\psi)_{\gamma_a}\left((D\iota_a)_{\hat{\gamma}}(v_1) \right), \ldots, (D\psi)_{\gamma_a}\left((D\iota_a)_{\hat{\gamma}}(v_N) \right) \right).
$$

By Lemma A.2, the sign of $(D\psi)_{\gamma_a}$ is given by the parity of $(\mu(x) + 1)\mu(y)$. Since $(D\psi)_{\gamma_a}$ maps the first vector of the basis to its opposite, the effect on the last N vectors is given by the parity of $(\mu(x) + 1)\mu(y) + 1$. By definition of the orientations involved, this implies that the basis in (A.8) is positive if and only if $(\mu(x) + 1)\mu(y) + 1$ is even, which shows the claim. $\qquad \square$

Remark A.4 In particular, if $\mu(x) = \mu(y) + 1$, i.e. if $\widehat{\mathcal{M}}(x, y, f)$ is zero-dimensional, then ψ is orientation preserving if and only if the following number is even:

$$(\mu(x) + 1)\mu(y) + 1 = (\mu(y) + 2)\mu(y) + 1 \equiv \mu(y)^2 + 1 \equiv \mu(y) + 1 \equiv \mu(x) \,.$$

In terms of oriented intersection numbers, this implies that

$$\#_{\mathrm{alg}}\widehat{\mathcal{M}}(x, y, f) = (-1)^{\mu(y)+1} \#_{\mathrm{alg}}\widehat{\mathcal{M}}(y, x, -f) = (-1)^{\mu(x)} \#_{\mathrm{alg}}\widehat{\mathcal{M}}(y, x, -f) \,.$$

We continue with the abovementioned investigation of the behaviour of Morse-theoretic gluing maps with respect to the chosen orientations on the trajectory spaces, starting with proper definitions of gluing maps.

Proposition A.5 ([Sch99, Lemma 4.1]) *Let $x, y \in \mathrm{Crit} f$ with $\mu(x) = \mu(y) + 1$ and let $V \subset \mathcal{W}^u(y)$ be an open subset. There exist $\rho_V > 0$ and a map*

$$G : \widehat{\mathcal{M}}(x, y) \times V \times [\rho_V, +\infty) \to \mathcal{W}^u(x)$$

with the following properties:

(i) G *is a smooth orientation-preserving embedding,*
(ii) *if $\{r_n\}_{n\in\mathbb{N}}$ is a sequence in $[\rho_V, +\infty)$ that diverges against $+\infty$, then $\{G(\hat{\gamma}, \gamma_0, r_n)\}_{n\in\mathbb{N}}$ will converge geometrically against $(\hat{\gamma}, \gamma_0)$ for all $\hat{\gamma} \in \widehat{\mathcal{M}}(x, y)$ and $\gamma_0 \in V$.*

We will call such a map a gluing map for negative half-trajectories.

See also [Sch05, Proposition 2.34] for a detailed proof of this existence result.

Proposition A.6 *Let $x_0, x_1 \in \mathrm{Crit} f$ with $\mu(x_0) = \mu(x_1) + 1$ and let $V \subset \mathcal{W}^s(x_0)$ be open. There exist $\rho_V > 0$ and a map*

$$G : V \times \widehat{\mathcal{M}}(x_0, x_1) \times [\rho_V, +\infty) \to \mathcal{W}^s(x_1)$$

with the following properties:

(i) G *is a smooth embedding which is orientation-preserving if and only if $\mu(x_0)$ is even,*
(ii) *if a sequence $\{r_n\}_{n\in\mathbb{N}}$ in $[\rho_V, +\infty)$ diverges against $+\infty$, then $\{G(\gamma_0, \hat{\gamma}, r_n)\}_{n\in\mathbb{N}}$ will converge geometrically against $(\gamma_0, \hat{\gamma})$ for all $\gamma_0 \in V$ and $\hat{\gamma} \in \widehat{\mathcal{M}}(x_0, x_1).$*

We will call such a map a gluing map for positive half-trajectories.

Proof Assume w.l.o.g. that $V = \mathcal{W}^s(x_0)$ and put $\rho_0 := \rho_{\mathcal{W}^s(x_0)}$. We will show the claim by identifying the stable manifolds of f with the unstable manifolds of $-f$ as we did earlier in this section. Afterwards, we apply part 1 of the proposition. For clarity's sake, we will write the domain of G_0 as $\mathcal{W}^s(x_0, f) \times \widehat{\mathcal{M}}(x_0, x_1, f) \times [\rho_0, +\infty)$.

Let $G' : \mathcal{M}(x_1, x_0, -f) \times \mathcal{W}^u(x_0, -f) \times [\rho_0, +\infty) \to \mathcal{W}^u(x_1, -f)$ be a gluing map for negative half-trajectories with respect to the Morse function $-f$. Define G as the map that makes the following diagram computative:

$$
\begin{array}{ccc}
\mathcal{W}^s(x_0, f) \times \widehat{\mathcal{M}}(x_0, x_1, f) \times [\rho_0, +\infty) & \xrightarrow{\ G\ } & \mathcal{W}^s(x_1, f) \\
\cong \downarrow \sigma & & \cong \downarrow \varphi_s \\
\widehat{\mathcal{M}}(x_1, x_0, -f) \times \mathcal{W}^u(x_0, -f) \times [\rho_0, +\infty) & \xrightarrow{\ G'\ } & \mathcal{W}^u(x_1, -f)
\end{array}
$$

where σ is the diffeomorphism defined as the composition of the transposition of the first two factors and the product of the maps φ_s and $\widehat{\psi}$ from above. Since G' is a smooth embedding and the vertical maps are diffeomorphisms, G is a smooth embedding as well. Moreover, one checks without difficulties that geometric convergence is preserved under the vertical diffeomorphisms, such that property (ii) of G' implies property (ii) for G. It remains to discuss the behaviour of G with respect to orientations.

By construction, the map $\varphi_s : \mathcal{W}^s(x_1, f) \to \mathcal{W}^u(x_1, -f)$ is orientation-preserving. Moreover, the bottom gluing map is orientation-preserving by Proposition A.5. The commutativity of the diagram thus implies that the sign of G is given by the sign of σ, which we compute as follows.

We first transpose $\widehat{\mathcal{M}}(x_0, x_1, f)$ and $\mathcal{W}^s(x_0, f)$. Since $\widehat{\mathcal{M}}(x_0, x_1, f)$ is zero-dimensional, this transposition is orientation-preserving. Afterwards, we apply φ_s to the factor $\mathcal{W}^s(x_0, f)$ and $\widehat{\psi}$ to $\widehat{\mathcal{M}}(x_0, x_1, f)$. As previously mentioned, φ_s is orientation-preserving and $\widehat{\psi}$ preserves orientations if and only if $\mu(x_0)$ is even, see Remark A.4. Hence, $\operatorname{sign} \sigma = (-1)^{\mu(x_0)}$, which shows that G has the desired properties. $\qquad\square$

The existence of gluing maps for positive half-trajectories follows immediately from the existence of gluing maps for negative half-trajectories by applying Proposition A.5 to the Morse function $-f$.

Proposition A.7 *Let $x \in \operatorname{Crit} f$ and $V_0 \subset \mathcal{W}^s(x)$ and $V_1 \subset \mathcal{W}^u(x)$ be open subsets. There exists $\rho_{V_0, V_1} > 0$ and a map*

$$
G : [\rho_{V_0, V_1}, +\infty) \times V_0 \times V_1 \to \mathcal{M}(f, g)
$$

with the following properties:

(i) *G is a smooth embedding,*

(ii) *if $\{r_n\}_{n \in \mathbb{N}}$ is a sequence in $[\rho_V, +\infty)$ that diverges against $+\infty$, then $\{G(r_n, \gamma_0, \gamma_1)\}_{n \in \mathbb{N}}$ will converge geometrically against (γ_0, γ_1) for all $\gamma_0 \in V_0$ and $\gamma_1 \in V_1$,*

(iii) *for all $\gamma_0 \in V_0$ and $\gamma_1 \in V_1$ the map $[\rho_{V_0, V_1}, +\infty) \to [0, +\infty), r \mapsto \ell(r, \gamma_0, \gamma_1)$, is strictly increasing, where $\ell(r, \gamma_0, \gamma_1)$ denotes the interval length of the finite-length trajectory $G(r, \gamma_0, \gamma_1)$.*

We will call such a map a gluing map for finite-length trajectories.

The existence of gluing maps for finite-length trajectories is discussed in [KM07, Chap. 18]. We want to investigate their behaviour with respect to orientations.

Proposition A.8 *Let $x \in \mathrm{Crit} f$, $V_0 \subset \mathcal{W}^s(x)$ and $V_1 \subset \mathcal{W}^u(x)$ be open subsets and let*

$$G : [\rho_{V_0, V_1}, +\infty) \times V_0 \times V_1 \to \mathcal{M}(f, g)$$

be a gluing map for finite-length trajectories. Then G is orientation-preserving if and only if $n + 1$ is even.

Proof Assume w.l.o.g. that $V_0 = \mathcal{W}^s(x)$ and $V_1 = \mathcal{W}^u(x)$ and put $\rho_0 := \rho_{\mathcal{W}^s(x), \mathcal{W}^u(x)}$. It suffices to show the claim at a fixed point $(\rho, \gamma_1, \gamma_2) \in [\rho_0, +\infty) \times \mathcal{W}^s(x) \times \mathcal{W}^u(x)$. Let γ_1 and γ_2 be the constant trajectories given by

$$\gamma_1(s) = x \ \forall s \in [0, +\infty) , \quad \gamma_2(s) = x \ \forall s \in (-\infty, 0] ,$$

and put $\ell : [\rho_0, +\infty) \to [0, +\infty)$, $\ell(r) := \ell(r, \gamma_1, \gamma_2)$, given as in Proposition A.7. By definition of a gluing map for finite-length trajectories, ℓ is smooth and strictly increasing, i.e. orientation-preserving. From this observation and the properties of gluing maps, it follows for the constant trajectories γ_1 and γ_2 and for all $\rho \in [\rho_0, +\infty)$ that $G(\rho, \gamma_1, \gamma_2)$ is the constant trajectory $[0, \ell(\rho)] \to M, t \mapsto x$.

One observes that the differential of G at $(\rho, \gamma_1, \gamma_2)$ is a map

$$DG_{(\rho, \gamma_1, \gamma_2)} : T_\rho[\rho_2, +\infty) \oplus T_{\gamma_1}\mathcal{W}^s(x) \oplus T_{\gamma_2}\mathcal{W}^u(x) \oplus \to T_{G(\rho, \gamma_1, \gamma_2)}\mathcal{M}(f, g)$$

that makes the following diagram commutative:

$$
\begin{array}{ccc}
T_\rho[\rho_0, +\infty) \oplus T_{\gamma_1}\mathcal{W}^s(x) \oplus T_{\gamma_2}\mathcal{W}^u(x) & \xrightarrow{DG_{(\rho, \gamma_1, \gamma_2)}} & T_{G(\rho, \gamma_1, \gamma_2)}\mathcal{M}(f, g) \\
\downarrow{\cong} & & \downarrow{\cong} \\
T_\rho[\rho_0, +\infty) \oplus T_x W^s(x) \oplus T_x W^u(x) & \xrightarrow{\cong} & T_{\ell(\rho)}[0, +\infty) \oplus T_x M ,
\end{array}
$$

Here, the bottom map is induced by $D\ell_\rho : T_\rho[\rho_2, +\infty) \to T_{l(\rho)}[0, +\infty)$ and the identification $T_x W^s(x) \oplus T_x W^u(x) \cong T_x M$, which are both orientation-preserving by definition of the orientation on $W^s(x)$ and the fact that ℓ is orientation-preserving. The left-hand vertical map is given by the differentials of the evaluation maps and is therefore orientation-preserving. The right-hand vertical map is given by the differential of the diffeomorphism $\mathcal{M}(f, g) \xrightarrow{\cong} [0, +\infty) \times M, (l, \gamma) \mapsto (l, \gamma(0))$, which has sign $(-1)^{n+1}$ by definition of the orientation on $\mathcal{M}(f, g)$. Consequently, the differential of the gluing map has sign $(-1)^{n+1}$ at $(\rho, \gamma_1, \gamma_2)$ and thus at every point since its domain is connected. $\qquad \square$

Remark A.9 In [Sch93, Chap. 3], the notion of coherent orientations is introduced to define orientations and algebraic counts of moduli spaces of unparametrized Morse trajectories. This approach has the big advantage that an orientation on the ambient manifold M is no longer required, which makes it possible to define Morse homology with integer coefficients on non-orientable manifolds. In [Sch93, Appendix B] it is shown that on an oriented closed manifold, both approaches to orienting trajectory spaces are indeed equivalent.

We conclude this section by a general statement on orientations of transverse intersections that is not directly connected to the above, but will be of great use in later sections.

For two finite-dimensional oriented vector spaces V_1 and V_2 we let *the direct sum orientation* on $V_1 \oplus V_2$ be given by demanding that a positive basis of V_1 followed by a positive basis of V_2 is a positive basis of $V_1 \oplus V_2$.

Theorem A.10 *Let M_1, M_2, P_1 and P_2 be smooth oriented manifolds and $N_1 \subset P_1$ and $N_2 \subset P_2$ be oriented submanifolds. Let $f_1 : M_1 \to P_1$ and $f_2 : M_2 \to P_2$ be smooth maps with $f_1 \pitchfork N_1$ and $f_2 \pitchfork N_2$. Let $S_1 := f_1^{-1}(N_1)$ and $S_2 := f_2^{-1}(N_2)$ be equipped with the orientations induced by the transverse intersections.*

If there are smooth diffeomorphisms $\varphi : M_1 \to M_2$ and $\psi : P_1 \to P_2$, such that

- *ψ restricts to a diffeomorphism $\psi|_{N_1} : N_1 \xrightarrow{\cong} N_2$,*

- *the following diagram commutes:*

$$\begin{CD} M_1 @>{\varphi}>{\cong}> M_2 \\ @V{f_1}VV @VV{f_2}V \\ P_1 @>{\psi}>{\cong}> P_2 , \end{CD}$$

then φ restricts to a diffeomorphism $\varphi|_{S_1} : S_1 \xrightarrow{\cong} S_2$ with

$$\mathrm{sign}(\varphi|_{S_1}) = \mathrm{sign}\varphi \cdot \mathrm{sign}\psi \cdot \mathrm{sign}(\psi|_{N_1}) .$$

Proof We assume w.l.o.g. that M_1 and M_2 are equipped with Riemannian metrics. Let $i \in \{1, 2\}$, $x \in S_i$ and let $N_x S_i \subset T_x M_i$ denote the normal space of S_i at x, such that

$$N_x S_i \oplus T_x S_i = T_x M_i . \tag{A.9}$$

Then $(Df_i)_x$ maps $N_x S_i$ isomorphically onto its image and since f_i is transverse to N_i, it holds that

$$(Df_i)_x[N_x S_i] \oplus T_{f_i(x)} N_i = T_{f_i(x)} P_i .$$

The orientations of N_i and P_i induce an orientation of $(Df_i)_x[N_x S_i]$, such that the direct sum orientation coincides with the given orientation of $T_x P_i$. This orientation of $(Df_i)_x[N_x S_i]$ induces an orientation of $N_x S_i$ by demanding that $(Df_i)_x$ is orientation-preserving. Finally, the orientations of $N_x S_i$ and of M_i induce an orientation of $T_x S_i$, such that the direct sum orientation in (A.9) coincides with the given orientation of $T_x M_i$.

To show the claim we need to prove that the orientation of $D\varphi_x[T_x S_1] = T_{\varphi(x)} S_2$ induced by φ coincides with the given orientation of S_2 if and only if $\text{sign}\varphi \cdot \text{sign}\psi \cdot \text{sign}(\psi|_{N_1})$ is positive. We start by applying the diffeomorphism φ to the decomposition (A.9), resulting in

$$D\varphi_x[N_x S_1] \oplus D\varphi_x[T_x S_1] = D\varphi_x[T_x M_1].$$

By definition, the orientation induced by φ on $D\varphi_x[T_x M_1] = T_{\varphi(x)} M_2$ coincides with the given orientation of M_2 if and only if φ is orientation-preserving. Thus, the sign of the orientation of the direct sum $D\varphi_x[N_x S_1] \oplus D\varphi_x[T_x S_1]$ induced by φ is given by $\text{sign}\varphi$.

To check the contribution of $D\varphi_x[T_x S_1]$ to this sign, we will first check the sign of $D\varphi_x[N_x S_1]$. By definition, the orientation of $D\varphi_x[N_x S_1]$ induced by φ is positive if and only if it induces a positive direct sum orientation

$$(Df_2)_{\varphi(x)}[D\varphi_x[N_x S_1]] \oplus T_{f_2(\varphi(x))} N_2 = T_{f_2(\varphi(x))} P_2.$$

By the commutativity of the above diagram this decomposition coincides with

$$D\psi_{f_1(x)}[(Df_1)_x[N_x S_1]] \oplus T_{f_2(\varphi(x))} N_2 = T_{f_2(\varphi(x))} P_2$$

Since ψ and its restriction to N_1 are diffeomorphisms, this equation coincides up to isomorphism with

$$D\psi_{f_1(x)}[(Df_1)_x[N_x S_1]] \oplus D(\psi|_{N_1})_{f_1(x)}[T_{f_1(x)} N_1] = D\psi_{f_1(x)}[T_{f_1(x)} P_1]$$

Hence, the orientation of $D\psi_{f_1(x)}[(Df_1)_x[N_x S_1]]$ is positive if and only if the orientations induced by $\psi|_{N_1}$ on $T_{f_2(\varphi(x))} N_2$ and by ψ on $T_{f_2(\varphi(x))} P_2$ are either both positive or both negative. In other words, the sign of the orientation of $(Df_2)_{\varphi_x}[D\varphi_x[N_x S_1]]$ is given by

$$\text{sign}\psi \cdot \text{sign}(\psi|_{N_1}).$$

Since the sign of the induced orientation on $D\varphi_x[N_x S_1] \oplus D\varphi_x[T_x S_1]$ is given by $\text{sign}\varphi$ and the sign of the induced orientation on $D\varphi_x[N_x S_1]$ is given by $\text{sign}\psi \cdot \text{sign}(\psi|_{N_1})$, the sign that $D\varphi_x$ induces on $T_x S_1$ is given by

$$\text{sign}\varphi \cdot \text{sign}\psi \cdot \text{sign}(\psi|_{N_1}),$$

which shows the claim.   □

A.2   Orientations of Perturbed Morse Trajectory Spaces

We will use the similarity between spaces of perturbed and unperturbed trajectories to equip spaces of perturbed Morse trajectories with orientations as well using the orientations on Morse trajectory spaces that we constructed in the previous section.

For $Y \in \mathfrak{X}_-(M)$ and $x \in \mathrm{Crit} f$ we define a map

$$\varphi_Y : W^-(x, Y) \to \mathcal{W}^u(x, f, g), \qquad \gamma \mapsto \left(s \mapsto \begin{cases} \gamma(s) & \text{if } s \leq -1, \\ \phi_{s+1}(\gamma(-1)) & \text{if } s \in (-1, 0], \end{cases} \right)$$

where ϕ denotes the negative gradient flow of f with respect to g.

Proposition A.11 *For every* $Y \in \mathfrak{X}_-(M)$, *the map* $\varphi_Y : W^-(x, Y) \to \mathcal{W}^u(x, f, g)$ *is a diffeomorphism of class* C^{n+1}.

Proof Using the unique existence of solutions of ordinary differential equations, one checks that the map φ_Y is indeed well-defined and bijective for every $Y \in \mathfrak{X}_-(T)$. In the following, we will identify the map with a composition of $(n + 1)$-times differentiable maps to show its differentiability properties. Consider the Banach manifold

$$\widetilde{\mathcal{P}}_-(x) := \left\{ \gamma \in H^1\left(\overline{\mathbb{R}}_{\leq -1}, M\right) \; \middle| \; \lim_{s \to -\infty} \gamma(s) = x \right\},$$

where $\overline{\mathbb{R}}_{\leq -1} := \{-\infty\} \cup (-\infty, -1]$, equipped with a smooth structure induced by the one on $\overline{\mathbb{R}}_{\leq 0}$. The restriction map

$$r : \mathcal{P}_-(x) \to \widetilde{\mathcal{P}}_-(x), \qquad \gamma \mapsto \gamma|_{\overline{\mathbb{R}}_{\leq -1}},$$

is obviously a smooth map of Banach manifolds. Moreover, again by the unique existence of solutions of ordinary differential equations, both of the restrictions

$$r|_{W^-(x,Y)} : W^-(x, Y) \to \widetilde{\mathcal{P}}_-(x), \qquad r|_{\mathcal{W}^u(x,f,g)} : \mathcal{W}^u(x, f, g) \to \widetilde{\mathcal{P}}_-(x),$$

are injective. Since by definition of $\mathfrak{X}_-(M)$, the time-dependent vector field Y vanishes in (s, x) if $s \leq -1$, the restriction $\gamma|_{(-\infty, -1]}$ satisfies the negative gradient flow equation of f with respect to g. It immediately follows that

$$\mathrm{im} r|_{W^-(x,Y)} = \mathrm{im} r|_{\mathcal{W}^u(x,f,g)} .$$

In terms of the restriction maps, we can therefore write φ_Y as the following well-defined composition:

$$\varphi_Y = \left(r|_{\mathcal{W}^u(x,f,g)}\right)^{-1} \circ r|_{W^-(x,Y)} .$$

Thus, φ_Y is a composition of a smooth map and a map of class C^{n+1} and is therefore itself of class C^{n+1}. Moreover, its inverse is given by

$$\varphi_Y^{-1} = \left(r|_{W(x,Y)}\right)^{-1} \circ r|_{\mathcal{W}^u(x,f,g)},$$

which is a map of class C^{n+1} by the very same argument. Therefore, φ_Y is a diffeomorphism of class C^{n+1}. □

For all $x \in \mathrm{Crit}\, f$ and $Y \in \mathfrak{X}_-(M)$, we equip $W^-(x, Y)$ with the unique orientation such that $\varphi_Y : W^-(x, Y) \overset{\cong}{\to} \mathcal{W}^u(x)$ is orientation-preserving.

We define a map for positive half-trajectories along the same lines. Let $x \in \mathrm{Crit}\, f$, $Y \in \mathfrak{X}_+(M)$ and define

$$\varphi_Y : W^+(x, Y) \to \mathcal{W}^s(x, f, g), \qquad \gamma \mapsto \left(s \mapsto \begin{cases} \phi_{s-1}(\gamma(1)) & \text{if } s \in [0, 1), \\ \gamma(s) & \text{if } s \geq 1. \end{cases}\right)$$

In the same way as in Proposition A.11, we show that in the positive case, the map φ_Y is a diffeomorphism of class C^{n+1} for any $Y \in \mathfrak{X}_+(M)$.

For all $x \in \mathrm{Crit}\, f$ and $Y \in \mathfrak{X}_+(M)$, we equip $W^+(x, Y)$ with the unique orientation for which $\varphi_Y : W^+(x, Y) \overset{\cong}{\to} \mathcal{W}^s(x)$ is orientation-preserving.

We recall from Proposition 2.19 that for $Y \in \mathfrak{X}_0(M)$ there is a diffeomorphism

$$\psi_Y : \mathcal{M}(Y) \to \mathcal{M}(f, g),$$

which is constructed in the spirit of the maps φ_Y.

For every $Y \in \mathfrak{X}_0(M)$, we equip $\mathcal{M}(Y)$ with the unique orientation such that the diffeomorphism $\psi_Y : \mathcal{M}(Y) \overset{\cong}{\to} \mathcal{M}(f, g)$ is orientation-preserving.

A.3   Ordering the Internal Edges of Ribbon Trees

Let $T \in \mathrm{BinTree}_d$ for some $d \geq 3$, such that $E_{int}(T) \neq \emptyset$. Given $e \in E_{int}(T) \cup \{e_0(T)\}$, the set $\{f \in E(T) \mid v_{in}(f) = v_{out}(e)\}$ has precisely two elements since T is a binary tree. We denote them by $f_1(e)$ and $f_2(e)$, which are uniquely defined by demanding that there exist $i, j \in \{1, 2, \ldots, d\}$ with $i < j$, such that

$$f_1(e) \in E(P_i(T)) \quad \text{and} \quad f_2(e) \in E(P_j(T)) \setminus E(P_i(T)),$$

and consider the maps

$$f_1, f_2 : E_{int}(T) \cup \{e_0(T)\} \to E(T), \qquad e \mapsto f_i(e) \text{ for } i \in \{1, 2\}.$$

Intuitively, drawing the trees as before, $f_1(e)$ denotes the left-hand edge and $f_2(e)$ the right-hand edge emanating from $v_{out}(e)$. For each $j \in \mathbb{N}$ for which it is well-defined we let $f_1^j(e) := (f_1 \circ f_1 \circ \cdots \circ f_1)(e)$ denote the j-fold iterate of f_1.

In addition we define a map

$$h : E(T) \setminus \{e_0(T)\} \to E(T) \,,$$

where $h(e)$ is the unique edge with $v_{\text{out}}(h(e)) = v_{\text{in}}(e)$ for every $e \in E(T) \setminus \{e_0(T)\}$. For every $j \in \mathbb{N}$ for which it is well-defined we denote the j-fold iterate of h by h^j.

We want to define a total ordering on $E_{int}(T)$. For this purpose, we will inductively label the edges as $\{g_1, g_2, \ldots, g_{d-2}\} = E_{int}(T)$ and define the ordering by demanding that g_i defines the ith element of $E_{int}(T)$. We first give a formal definition of the ordering procedure and afterwards describe it in a more informal and intuitive way.

Let $g_0 \in E(T)$ be the unique edge with $v_{\text{out}}(g_0) = v_{\text{in}}(e_1(T))$. We put

$$g_1 := \begin{cases} g_0 & \text{if } g_0 \in E_{int}(T) \,, \\ f_2(g_0) & \text{if } g_0 \notin E_{int}(T) \,. \end{cases}$$

Since $d \geq 3$, it holds that $g_1 \in E_{int}(T)$.

Assume that we have already defined the labelled edges $\{g_1, g_2, \ldots, g_{i-1}\}$ for some $i \in \{2, 3, \ldots, d-2\}$.

- If $f_1(g_{i-1}) \in E_{int}(T) \setminus \{g_1, g_2, \ldots, g_{i-1}\}$, then we put

$$g_i := f_1^k(g_{i-1}) \,,$$

  where $k = \max\{j \in \mathbb{N} \mid f_1^j(g_{i-1}) \in E_{int}(T)\}$.
- Assume that $f_1(g_{i-1}) \in E_{ext}(T) \cup \{g_1, \ldots, g_{i-1}\}$.

  - If $f_2(g_{i-1}) \in E_{int}(T)$, then put $g_i := f_2(g_{i-1})$.
  - If $f_2(g_{i-1}) \in E_{ext}(T)$, then put $g_i := h^k(g_{i-1})$, where

$$k = \min\{j \in \mathbb{N} \mid h^j(g_{i-1}) \in E_{int}(T) \setminus \{g_1, \ldots, g_{i-1}\}\} \,.$$

Less formally speaking, the above ordering can be described in simpler words. We first consider the edge for which $e_1(T)$ is the (left-hand) edge attached to its outgoing vertex. If this edge is internal, we define it as the first edge of the ordering.

If this edge is external, it is necessarily given by $e_0(T)$. Since T is binary and has at least three leaves, it follows that $f_2(e_0(T))$ is internal and we define $f_2(e_0(T))$ as the first edge. Intuitively, if the left-hand subtree emanating from $v_{\text{out}}(e_0(T))$ consists of a single external edge, then we define the edge of the right-hand subtree that is closest to the root of T as the first one.

Assuming that we have labelled the first $i - 1$ edges, we want to define the ith one, for which we have to distinguish several cases.

If the left-hand edge emanating from the $(i - 1)$-st edge, call it e, is internal and has not been labelled yet, then we check if the left-hand edge emanating from e has

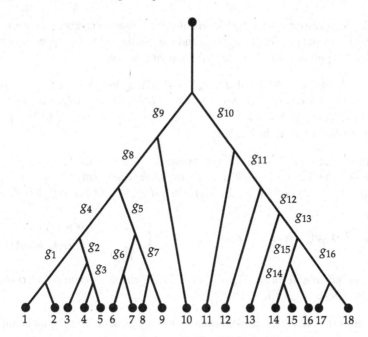

Fig. A.1 An example of the canonical ordering of the internal edges of a binary tree with $d = 18$ leaves

the very same property. We iterate this procedure and follow the unlabelled left-hand edges until we have reached the bottom of the tree. The last internal edge that we obtain is this way is then defined as the ith edge of the ordering.

   If the left-hand edge emanating from the $(i - 1)$-st edge is external or already labelled, then we consider the right-hand edge emanating from the $(i - 1)$-st edge. If that edge is internal, then we define it as g_i. If not, then we "backtrack" along the tree. We check if the edge "on top" of g_{i-1}, i.e. the edge g whose outgoing vertex is attached to the incoming vertex of g_{i-1}, is labelled or not. If g is unlabelled we let it be the ith edge. If g is labelled, we continue with the edge whose outgoing vertex coincides with $v_{in}(g)$ and check if it is labelled. We iterate this procedure until we arrive at an unlabelled edge. This edge is then defined to be the ith one.

Definition A.12 Let $T \in \mathrm{BinTree}_d$ for some $d \geq 3$. We call the ordering of $E_{int}(T)$ which is obtained by the above procedure *the canonical ordering of $E_{int}(T)$*.

See Fig. A.1 for an example of the canonical ordering of a binary tree. The canonical ordering is described as an implementation of a depth-first search algorithm for binary trees. Thus we refrain from a proof of the following statement.

Proposition A.13 *The canonical ordering is a well-defined total order of $E_{int}(T)$ for every $T \in \mathrm{BinTree}_d$ with $d \geq 3$.*

We will next investigate how the canonical orientations behave with respect to the decomposition operation of binary trees and the splitting operation along an internal edge. For this purpose, we introduce the following notion.

Definition A.14 Let $d \in \mathbb{N}$ with $d \geq 2, T \in \text{BinTree}_d$ and $e \in E(T) \setminus \{e_0(T)\}$. The edge e is called *left-handed* if there exists $g \in E(T)$ with $e = f_1(g)$ and *right-handed* if there exists $h \in E(T)$ with $e = f_2(h)$. Note that every element of $E(T) \setminus \{e_0(T)\}$ is either left-handed or right-handed.

Proposition A.15 Let $T \in \text{BinTree}_d$ for some $d \geq 3$ and let $E_{int}(T) = \{g_1, g_2, \ldots, g_{d-2}\}$, where the edges are labelled according to the canonical ordering of $E_{int}(T)$. Let $i \in \{1, 2, \ldots, d\}$ and $l \in \{1, 2, \ldots, d - i\}$. If $e \in E_{int}(T)$ is of type (i, l), then

$$E_{int}(T_2^e) = \{g_i, g_{i+1}, \ldots, g_{i+l-2}\}, \qquad e = \begin{cases} g_{i+l-1} & \text{if } e \text{ is left-handed,} \\ g_{i-1} & \text{if } e \text{ is right-handed,} \end{cases}$$

and the canonical ordering of $E_{int}(T_2^e)$ coincides with the ordering which is induced by the ordering of $E_{int}(T)$.

Proof This is derived from the definition of the canonical ordering by an elementary line of argument. Thus, we omit the details. □

Theorem A.16 Let $T \in \text{RTree}_d$ for some $d \geq 3$, $i \in \{1, 2, \ldots, d\}$, $l \in \{1, 2, \ldots, d - i\}$ and $e \in E_{int}(T)$ be of type (i, l). Let $E_{int}, E_{int}(T_1^e), E_{int}(T_2^e)$ be equipped with the canonical orderings. Then the sign of the permutation

$$\tau_e : \{e\} \times E_{int}(T_1^e) \times E_{int}(T_2^e) \to E_{int}(T)$$

is given by

$$\text{sign}\tau_e = \begin{cases} (-1)^{(d-i)l+d-1} & \text{if } e \text{ is left-handed,} \\ (-1)^{(d-i)l+d} & \text{if } e \text{ is right-handed.} \end{cases}$$

Proof If e is of type (i, l), then by Proposition A.15 it will hold that

$$E_{int}(T_2^e) = \{g_i, g_{i+1}, \ldots, g_{i+l-2}\},$$

$$E_{int}(T_1^e) = \begin{cases} \{g_1, \ldots, g_{i-2}, g_{i+l-1}, \ldots, g_{d-2}\} & \text{if } e = g_{i-1}, \\ \{g_1, \ldots, g_{i-1}, g_{i+l}, \ldots, g_{d-2}\} & \text{if } e = g_{i+l-1}. \end{cases}$$

If $e = g_{i+l-1}$, i.e. if e is left-handed, then $\tau_e = \tau_2 \circ \tau_1$, where τ_1 permutes e past $(g_1, \ldots, g_{i-1})$ and τ_2 permutes $(g_i, \ldots, g_{i+l-2})$ past $(g_{i+l-1}, \ldots, g_{d-2})$ with respect to the orderings. Hence, if $e = g_{i+l-1}$, the sign of τ_e is given by the parity of

$$i - 1 + (l - 1)(d - 2 - (i + l - 2)) = i - 1 + (l - 1)(d - i - l)$$

$$\equiv i - 1 + (l - 1)(d - i) \equiv d - 1 + l(d - i).$$

Analogously, if $e = g_{i-1}$, i.e. if e is right-handed, then τ_e permutes e past $(g_1, \ldots, g_{i-2})$ and $(g_i, \ldots, g_{i+l-2})$ past $(g_{i+l-1}, \ldots, g_{d-2})$ with respect to the orderings, such that $\mathrm{sign}\,\tau_e$ is given by the parity of

$$i - 2 + (l - 1)(d - 2 - (i + l - 2)) \equiv (d - i)l + d .$$

$\square$

So far, we have defined orderings of internal edges of *binary* trees. We will generalize canonical orderings to edges of arbitrary ribbon trees starting from the binary case. One checks that for every $T \in \mathrm{RTree}_d$ there exist $\tilde{T} \in \mathrm{BinTree}_d$ and $F \subset E_{int}\left(\tilde{T}\right)$, such that

$$T \cong \tilde{T}/F .$$

Definition A.17 Let $d \geq 3$ and $T \in \mathrm{RTree}_d$. Let $\tilde{T} \in \mathrm{BinTree}_d$ and $F \subset E_{int}\left(\tilde{T}\right)$ with $T \cong \tilde{T}/F$. Then

$$E_{int}(T) = E_{int}\left(\tilde{T}\right) \setminus F \tag{A.10}$$

and we define *the canonical ordering of* $E_{int}(T)$ as the ordering which is induced by the canonical ordering of $E_{int}\left(\tilde{T}\right)$ and the identification (A.10).

By elementary arguments from graph theory, one shows that the canonical ordering of internal edges of ribbon trees is well-defined, i.e. that for every $T \in \mathrm{RTree}_d$ there exists a pair $(\tilde{T}, F)$ as in the previous definition and that the ordering is independent of the choice of $(\tilde{T}, F)$.

The following quantity will be used in Sect. A.4.

Definition A.18 Let $d \in \mathbb{N}$ with $d \geq 2$ and $T \in \mathrm{BinTree}_d$. The *dexterity of* T is the number $r(T) \in \mathbb{N}_0$ defined by

$$r(T) = |\{g \in E_{int}(T) \mid g \text{ is right-handed}\}| .$$

Proposition A.19 *Let* $d \in \mathbb{N}$ *with* $d \geq 3$.

(a) Let $T_1, T_2 \in \mathrm{BinTree}_d$, $e_1 \in E_{int}(T_1)$ *and* $e_2 \in E_{int}(T_2)$ *be given such that* $T_1/e_1 \cong T_2/e_2$. *Then*
$$|r(T_1) - r(T_2)| = 1 .$$

(b) Let $T \in \mathrm{RTree}_d$ *and* $e \in E_{int}(T)$. *Then*

$$r(T_1^e) + r(T_2^e) = \begin{cases} r(T) & \text{if } e \text{ is left-handed,} \\ r(T) - 1 & \text{if } e \text{ is right-handed.} \end{cases}$$

Proof (a) Apparently, precisely one of the two edges e_1 and e_2 is right-handed. Consequently, one of the two trees T_1 and T_2 has an additional right-handed edge.

(b) This is obvious.

□

A.4 Orientations on Moduli Spaces of Perturbed Morse Ribbon Trees

Throughout this section, we assume that $E_{int}(T)$ is equipped with the canonical ordering for all $T \in \mathrm{RTree}_d$ and $d \geq 3$. Assume that such a T is chosen and that we consider a product of the form $\prod_{e \in E_{int}(T)} Q_e$, where Q_e is an oriented manifold for every $e \in E_{int}(T)$. Throughout this section, we assume that such a product is oriented with the product orientation, given by

$$\prod_{e \in E_{int}(T)} Q_e = Q_{g_1} \times Q_{g_2} \times \cdots \times Q_{g_{k(T)}} \,,$$

where the elements of $E_{int}(T) = \{g_1, g_2, \ldots, g_{k(T)}\}$ are labelled according to the canonical ordering.

Let $d \geq 2$, $T \in \mathrm{RTree}_d$ and $\mathbf{Y} \in \mathfrak{X}(T)$. As a first step towards orientations on the spaces $\mathcal{A}_{\mathbf{Y}}^d(x_0, x_1, \ldots, x_d, T)$, we want to equip the spaces

$$\mathcal{M}_{\mathbf{Y}}^d(x_0, x_1, \ldots, x_d, T) \,, \quad \text{for } x_0, x_1, \ldots, x_d \in \mathrm{Crit} f \,,$$

from Sect. 6.2 with orientations. Recall that in shorthand notation

$$\mathcal{M}_{\mathbf{Y}}^d(x_0, x_1, \ldots, x_d, T) = \big\{ (\gamma_0, (l_e, \gamma_e)_{e \in E_{int}(T)}, \gamma_1, \ldots, \gamma_d) \mid \gamma_0 \in W^-(x_0, Y_0((l_e)_e)),$$
$$l_e > 0, \ (l_e, \gamma_e) \in \mathcal{M}(Y_e((l_f)_{f \neq e})) \ \forall e, \ \gamma_i \in W^+(x_i, Y_i((l_e)_e)) \ \forall i \big\} \,,$$

with $\mathbf{Y} = (Y_0, (Y_e)_{e \in E_{int}(T)}, Y_1, \ldots, Y_d)$. As we discussed in Sect. 6.2, the space $\mathcal{M}_{\mathbf{Y}}^d(x_0, x_1, \ldots, x_d, T)$ is a manifold of class C^{n+1} and one checks without difficulties that the following map is a diffeomorphism of manifolds of class C^{n+1}:

$$F_T : \mathcal{M}_{\mathbf{Y}}^d(x_0, x_1, \ldots, x_d, T) \overset{\cong}{\to} W^u(x_0) \times \prod_{e \in E_{int}(T)} \mathcal{M}(f, g) \times W^s(x_1) \times \cdots \times W^s(x_d) \,,$$

$$(\gamma_0, (l_e, \gamma_e)_e, \gamma_1, \ldots, \gamma_d) \mapsto$$

$$\left(\varphi_{Y_0((l_e)_e)}(\gamma_0), \left(\psi_{Y_e((l_f)_{f \neq e})}(l_e, \gamma_e) \right)_{e \in E_{int}(T)}, \varphi_{Y_1((l_e)_e)}(\gamma_1), \ldots, \varphi_{Y_d((l_e)_e)}(\gamma_d) \right) \,.$$

For all $d \geq 2$, $T \in \mathrm{RTree}_d$ and $x_0, x_1, \ldots, x_d \in \mathrm{Crit} f$, we equip $\mathcal{M}_{\mathbf{Y}}^d(x_0, x_1, \ldots, x_d, T)$ with the unique orientation with which F_T becomes an orientation-preserving diffeomorphism.

As a second step towards the construction of orientations on moduli spaces of type $\mathcal{A}_{\mathbf{Y}}^d(x_0, x_1, \ldots, x_d, T)$, we investigate the orientability of the T-diagonals.

Proposition A.20 *Let* $d \geq 2$. *For every* $T \in \mathrm{RTree}_d$ *the canonical ordering of* $E_{int}(T)$ *induces a diffeomorphism*

$$M^{1+k(T)} \overset{\cong}{\longrightarrow} \Delta_T .$$

Proof We identify $M^{k(T)}$ with $M^{E_{int}(T)}$ and $M^{2k(T)}$ with $\left(M^2\right)^{E_{int}(T)}$ according to the canonical ordering and define a map

$$M^{1+k(T)} \rightarrow M^{1+2k(T)+d} , \quad \left(x_0, (x_e)_{e \in E_{int}(T)}\right) \mapsto \left(x_0, \left(q_{in}^e, x_e\right)_{e \in E_{int}(T)}, q_1, q_2, \ldots, q_d\right) ,$$

where

$$q_{in}^e := \begin{cases} x_0 & \text{if } v_{in}(e) = v_{out}(e_0(T)) , \\ x_g & \text{if } v_{in}(e) = v_{out}(g) \text{ for some } g \in E_{int}(T) , \end{cases}$$

$$q_i := \begin{cases} x_0 & \text{if } v_{in}(e_i(T)) = v_{out}(e_0(T)) , \\ x_g & \text{if } v_{in}(e_i(T)) = v_{out}(g) \text{ for some } g \in E_{int}(T) , \end{cases}$$

for all $e \in E_{int}(T)$ and $i \in \{1, 2, \ldots, d\}$. One checks without difficulties that this map is well-defined and a diffeomorphism onto its image. Moreover, it follows from the description of Δ_T in Remark 4.12 that the image of this map is Δ_T.   □

For all $d \geq 2$ *and* $T \in \mathrm{RTree}_d$, *we equip* Δ_T *with the unique orientation which makes the diffeomorphism* $M^{1+k(T)} \cong \Delta_T$ *from Proposition A.20 orientation-preserving.*

Next we will use the last results to construct the desired orientations on moduli spaces of perturbed Morse ribbon trees. Let $T \in \mathrm{RTree}_d$ and $x_0, x_1, \ldots, x_d \in \mathrm{Crit} f$. As we have described in Sect. 6.2, the space $\mathcal{A}_{\mathbf{Y}}^d(x_0, x_1, \ldots, x_d, T)$ is a transverse intersection of manifolds of class C^{n+1}, namely

$$\mathcal{A}_{\mathbf{Y}}^d(x_0, x_1, \ldots, x_d, T) = \underline{E}_{\mathbf{Y}}^{-1}(\Delta_T) ,$$

where $\underline{E}_{\mathbf{Y}} : \mathcal{M}_{\mathbf{Y}}^d(x_0, x_1, \ldots, x_d, T) \rightarrow M^{1+2k(T)+d}$ is the endpoint evaluation map defined in (6.2). Hence, if $\mathbf{Y}$ is a regular perturbation datum, the space $\mathcal{A}_{\mathbf{Y}}^d(x_0, x_1, \ldots, x_d, T)$ is the transverse intersection of an oriented manifold with an oriented submanifold of an oriented manifold.

By standard results from differential topology the given orientations induce an orientation on $\mathcal{A}_{\mathbf{Y}}^d(x_0, x_1, \ldots, x_d, T)$ in this case, see [GP74, Chap. 3] or [BH04, Sect. 5.6].

For all $x_0, x_1, \ldots, x_d \in \mathrm{Crit} f$ *and* $T \in \mathrm{RTree}_d$ *let* $\mathcal{A}_{\mathbf{Y}}^d(x_0, x_1, \ldots, x_d, T)$ *be equipped with the orientation induced by its description as a transverse intersection of oriented manifolds.*

In the following, we will focus on the zero-dimensional case and briefly repeat the definition of oriented intersection numbers, following the approach of [GP74, Chap. 3].

Let M and P be finite-dimensional smooth oriented manifolds and $N \subset P$ be an oriented submanifold. Put $m := \dim M$, $n := \dim N$ and $p := \dim P$ and assume that

$$m + n = p \, .$$

Let $f : M \to P$ be a smooth map transverse to N, which we write as $f \pitchfork N$. Then $S := f^{-1}(N)$ is a zero-dimensional manifold and the orientations on M, N and P induce an orientation on S, which we express as a map $\epsilon_S : S \to \{-1; 1\}$, where $\epsilon_S(x) = +1$ if and only if $x \in S$ satisfies the following condition:

If $(v_1, v_2, \ldots, v_m)$ is a positive basis of $T_x M$ and $(w_1, w_2, \ldots, w_n)$ is a positive basis of $T_{f(x)} N$, then

$$(df_x[v_1], df_x[v_2], \ldots, df_x[v_m], w_1, w_2, \ldots, w_n)$$

is a positive basis of $T_{f(x)} P$.

The *oriented intersection number* or *algebraic count* of S is then defined as

$$\#_{\mathrm{or}} S := \sum_{x \in S} \epsilon_S(x) \in \mathbb{Z} \cup \{-\infty, +\infty\} \, .$$

It is an elementary fact that this number is independent of the choices of bases, hence well-defined. See [GP74, Chap. 3], [Hir76, Sect. 5.2] or [BH04, Sect. 5.6] for further details and generalities on this construction.

We make this definition of oriented intersection numbers explicit for zero-dimensional spaces $\mathcal{A}_{\mathbf{Y}}^d(x_0, x_1, \ldots, x_d, T)$. Let $\mathbf{Y}$ be a regular perturbation datum, $d \geq 2$, $T \in \mathrm{RTree}_d$ and $x_0, x_1, \ldots, x_d \in \mathrm{Crit} f$ such that $\mathcal{A}_{\mathbf{Y}}^d(x_0, x_1, \ldots, x_d, T)$ is a zero-dimensional manifold and put

$$\epsilon_T := \epsilon_{\mathcal{A}_{\mathbf{Y}}^d(x_0, x_1, \ldots, x_d, T)} : \mathcal{A}_{\mathbf{Y}}^d(x_0, x_1, \ldots, x_d, T) \to \{-1, 1\} \, .$$

Let $\underline{\gamma} \in \mathcal{A}_{\mathbf{Y}}^d(x_0, x_1, \ldots, x_d, T)$. For suitable numbers $N_1, N_2 \in \mathbb{N}_0$, we let $(b_1, b_2, \ldots, b_{N_1})$ be a positive basis of $T_{\underline{\gamma}} \mathcal{M}_{\mathbf{Y}}^d(x_0, x_1, \ldots, x_d, T)$ and $(\beta_1, \beta_2, \ldots, \beta_{N_2})$ be a positive basis of $T_{\underline{E}_{\mathbf{Y}}(\underline{\gamma})} \Delta_T$. Then $\epsilon_T \left(\underline{\gamma} \right) = +1$ if and only if

$$\left((D\underline{E}_{\mathbf{Y}})_{\underline{\gamma}} [b_1], (D\underline{E}_{\mathbf{Y}})_{\underline{\gamma}} [b_2], \ldots, (D\underline{E}_{\mathbf{Y}})_{\underline{\gamma}} [b_{N_1}], \beta_1, \beta_2, \ldots, \beta_{N_2} \right) \qquad \text{(A.11)}$$

is a *positive* basis of $T_{\underline{E}_{\mathbf{Y}}(\underline{\gamma})} M^{1+2k(T)+d}$.

Finally, we are in a position to define the coefficients of the higher order multiplications in terms of oriented intersection numbers.

Definition A.21 Let $d \in \mathbb{N}$ with $d \geq 2$ and $x_0, x_1, \ldots, x_d \in \mathrm{Crit} f$ satisfy

$$\mu(x_0) = \sum_{i=1}^{d} \mu(x_i) + 2 - d .$$ (A.12)

We define $a_{\mathbf{Y}}^d(x_0, x_1, \ldots, x_d) \in \mathbb{Z}$ by

$$a_{\mathbf{Y}}^d(x_0, x_1, \ldots, x_d) = \sum_{T \in \mathrm{RTree}_d} (-1)^{\sigma(x_0, x_1, \ldots, x_d) + r(T)} \#_{\mathrm{or}} \mathcal{A}_{\mathbf{Y}}^d(x_0, x_1, \ldots, x_d, T) ,$$

where $r(T)$ denotes the dexterity of T for every $T \in \mathrm{RTree}_d$.

Note that this definition of the coefficients will coincide with the one from Definition 6.3 if we put

$$\#_{\mathrm{alg}} \mathcal{A}_{\mathbf{Y}}^d(x_0, x_1, \ldots, x_d, T) := (-1)^{r(T)} \#_{\mathrm{or}} \mathcal{A}_{\mathbf{Y}}^d(x_0, x_1, \ldots, x_d, T)$$

for every $T \in \mathrm{RTree}_d$. Moreover, since for critical points satisfying (A.12), it holds that $\mathcal{A}_{\mathbf{Y}}^d(x_0, x_1, \ldots, x_d, T) = \varnothing$ whenever $T \notin \mathrm{BinTree}_d$, it follows that

$$a_{\mathbf{Y}}^d(x_0, x_1, \ldots, x_d) = \sum_{T \in \mathrm{BinTree}_d} (-1)^{\sigma(x_0, x_1, \ldots, x_d) + r(T)} \#_{\mathrm{or}} \mathcal{A}_{\mathbf{Y}}^d(x_0, x_1, \ldots, x_d, T) .$$

In analogy with the unperturbed case, it is possible to define gluing maps for spaces of perturbed Morse trajectories and more generally for the spaces discussed in Chap. 3, in particular for spaces of perturbed Morse ribbon trees. Since the technical effort behind carrying out the details of the constructions is disproportionate, we will not carry out the analysis of gluing maps for spaces of perturbed Morse trajectories. We note that, as in the unperturbed case, one reduces the question of existence of gluing maps to an application of the Banach Fixed-Point Theorem. Instead of carrying out the details, we will work with the following slightly vague definition that we already give for later purposes.

Definition A.22 Let $\mathcal{A}$ and $\mathcal{B}$ be differentiable manifolds whose elements are families of perturbed Morse trajectories of all three types (negative semi-infinite, positive semi-infinite and finite-length). We call a map $G : [\rho_0, +\infty) \times \mathcal{B} \to \mathcal{A}$ a *geometric gluing map* if it has the following properties:

- G is a differentiable embedding,
- if $\{r_n\}_{n \in \mathbb{N}}$ is a sequence in $[\rho_0, +\infty)$ that diverges to $+\infty$, then the sequence $\left\{ G\left(r_n, \underline{\gamma}\right) \right\}_{n \in \mathbb{N}}$ in $\mathcal{A}$ will converge geometrically against $\underline{\gamma}$ for every $\underline{\gamma} \in \mathcal{B}$.

Here, we define geometric convergence by demanding that all component sequences of perturbed Morse trajectories converge or converge geometrically.

A.5 Proof of Theorem 6.28

Throughout this section, we let $\mathbf{Y}$ be an admissible perturbation datum. We further let $d \geq 2$ and $x_0, x_1, \ldots, x_d \in \operatorname{Crit} f$ be chosen such that

$$\mu(x_0) = \sum_{q=1}^{d} \mu(x_q) + 3 - d \,, \tag{A.13}$$

i.e. such that $\mathcal{A}_{\mathbf{Y}}^d(x_0, x_1, \ldots, x_d, T)$ is a one-dimensional manifold for every $T \in \operatorname{BinTree}_d$. We will further make frequent use of the following helpful observation whose proof can e.g. be found in [GP74, p. 101].

Proposition A.23 *Let P be a smooth oriented manifold with boundary, Q be a smooth oriented manifold and $N \subset Q$ be a smooth oriented submanifold, both without boundary. Let $f : P \to Q$ be of class C^1 and assume that both f and $f|_{\partial M}$ are transverse to N, such that $S := f^{-1}(N)$ is a manifold with boundary. Equip S with the orientation induced by the transverse intersection.*

The boundary orientation on ∂S coincides with the orientation induced by the transverse intersection of $f|_{\partial M}$ with N if and only if

$$\operatorname{codim}_Q N$$

is even.

In the previous section, we have defined orientations on the spaces $\mathcal{A}_{\mathbf{Y}}^d(x_0, x_1, \ldots, x_d, T)$ for fixed $T \in \operatorname{RTree}_d$. If the critical points are chosen to satisfy (A.13), these spaces are at most one-dimensional. In the one-dimensional case their orientations induce orientations of their compactifications $\overline{\mathcal{A}}_{\mathbf{Y}}^d(x_0, x_1, \ldots, x_d, T)$.

We will consider the space $\overline{\mathcal{A}}_{\mathbf{Y}}^d(x_0, x_1, \ldots, x_d, T)$ as equipped with this extended orientation for every $T \in \operatorname{RTree}_d$.

The space $\overline{\mathcal{A}}_{\mathbf{Y}}^d(x_0, x_1, \ldots, x_d)$ is defined in Sect. 6.6 as a quotient space of the disjoint union of the $\overline{\mathcal{A}}_{\mathbf{Y}}^d(x_0, x_1, \ldots, x_d, T)$ over all binary trees T.

To prove Theorem 6.28, we need to define an orientation on the space $\overline{\mathcal{A}}_{\mathbf{Y}}^d(x_0, x_1, \ldots, x_d)$. It is a natural question if the orientations on the spaces $\overline{\mathcal{A}}_{\mathbf{Y}}^d(x_0, x_1, \ldots, x_d, T)$ induce an orientation on $\overline{\mathcal{A}}_{\mathbf{Y}}^d(x_0, x_1, \ldots, x_d)$ and it is precisely at this point that we will finally use the properties of the canonical orderings of sets of internal edges.

Lemma A.24 *Let $T_1, T_2 \in \operatorname{BinTree}_d$, $T \in \operatorname{RTree}_d$, $e_1 \in E_{int}(T_1)$ and $e_2 \in E_{int}(T_2)$, such that*

$$T_1/e_1 = T = T_2/e_2 \,.$$

The two boundary orientations on $\mathcal{A}_{\mathbf{Y}}^d(x_0, x_1, \ldots, x_d, T)$, which are obtained by identifying $\mathcal{A}_{\mathbf{Y}}^d(x_0, x_1, \ldots, x_d, T)$ as a subset of the boundary of

$$both \quad \overline{\mathcal{A}}_{\mathbf{Y}}^d(x_0, x_1, \ldots, x_d, T_1) \quad and \quad \overline{\mathcal{A}}_{\mathbf{Y}}^d(x_0, x_1, \ldots, x_d, T_2) \, ,$$

coincide.

Because of its technicalities, we will only outline the proof of Lemma A.24.

Sketch of proof Apparently, precisely one of the two edges e_1 and e_2 is left-handed and we assume w.l.o.g. that e_1 is the left-handed one. In the following, we identify

$$E_{int}(T) = E_{int}(T_1) \setminus \{e_1\} = E_{int}(T_2) \setminus \{e_2\} \, .$$

Let f_1, f_2, f_0, $g \in E_{int}(T)$ be given by demanding that f_1 is left-handed and that in T_1, it holds that

$$v_{in}(e_1) = v_{out}(f_0), \quad v_{out}(e_1) = v_{in}(f_1) = v_{in}(f_2), \quad v_{in}(e_1) = v_{in}(g) \, ,$$

such that g is right-handed in T_1. One checks from the definitions of the trees that g is then left-handed in T_2 and that in T_2, it holds that

$$v_{in}(e_2) = v_{out}(f_0), \quad v_{out}(e_2) = v_{in}(g) = v_{in}(f_2), \quad v_{in}(e_2) = v_{in}(f_1) \, .$$

Moreover, all other vertex identifications of T_1 coincide with identifications of T_2 and vice versa. Thus, one checks from these equalities that one gets a diffeomorphism

$$\varphi : M^{1+2k(T_1)+d} \to M^{1+2(T_2)+d}$$

with $\varphi(\Delta_{T_1}) = \Delta_{T_2}$ as follows: writing elements of $M^{1+2k(T_2)+d}$ as

$$(q_0, (q_e^1, q_e^2)_{e \in E_{int}(T_1)}, q_1, \ldots, q_d) \, ,$$

then we need to interchange the q_g^2-component and the $q_{f_1}^2$-component and afterwards permute the pairs (q_e^1, q_e^2) according to the orderings of the edges, where $(q_{e_1}^1, q_{e_1}^2)$ takes the role of the component associated with e_2. Since the latter permutation only permutes the order of even-dimensional manifolds, it is orientation-preserving. Moreover, one checks that transposing the two components is orientation-preserving if and only if n is even, independent of the orderings of the edges.

Taking a closer look at orientations of T-diagonals one checks that the sign of $\varphi|_{\Delta_{T_1}}$ coincides with the sign of the permutation $M^{1+k(T_1)} \to M^{1+k(T_2)}$ that interchanges the factors associated with f_1 and g and leaves all other factors invariant. Consequently,

$$\mathrm{sign}\varphi = \mathrm{sign}(\varphi|_{\Delta_{T_1}}) = (-1)^n \, .$$

Let $j \in \{1, 2, \ldots, d-2\}$ be given such that e_1 is the jth internal edge of T_1 with respect to the canonical ordering. Then it follows from the definition of the canonical orderings that e_2 is the jth internal edge of T_2 as well and that the map

$$E_{int}(T_1) \to E_{int}(T_2), \quad e \mapsto \begin{cases} e & \text{if } e \neq e_1, \\ e_2 & \text{if } e = e_1, \end{cases}$$

preserves the canonical orderings. By definition of the moduli spaces, one derives that this map induces a diffeomorphism

$$\mathcal{M}_{\mathbf{Y}}^d(x_0, x_1, \ldots, x_d, T_1) \to \mathcal{M}_{\mathbf{Y}}^d(x_0, x_1, \ldots, x_d, T_2)$$

of class C^{n+1} which is orientation-preserving. Since $\mathcal{A}_{\mathbf{Y}}^d(x_0, x_1, \ldots, x_d, T_i)$ is for $i \in \{1, 2\}$ given as the transverse intersection of $\mathcal{M}_{\mathbf{Y}}(x_0, x_1, \ldots, x_d, T_i)$ with Δ_{T_i}, it follows from Theorem A.10 that the identity induces an diffeomorphism

$$\mathcal{A}_{\mathbf{Y}}^d(x_0, x_1, \ldots, x_d, T_1) \to \mathcal{A}_{\mathbf{Y}}^d(x_0, x_1, \ldots, x_d, T_2)$$

whose sign is given by

$$\text{sign}\varphi \cdot \text{sign}(\varphi|_{\Delta_{T_i}}) = (-1)^n \cdot (-1)^n = 1 \,,$$

i.e. the diffeomorphism is orientation-preserving. The claim then follows from passing to the boundary orientations in both cases. $\qquad\square$

For every $T \in \text{BinTree}_d$ we let $\widetilde{\mathcal{A}}_{\mathbf{Y}}^d(x_0, x_1, \ldots, x_d, T)$ denote the oriented manifold diffeomorphic to $\overline{\mathcal{A}}_{\mathbf{Y}}^d(x_0, x_1, \ldots, x_d, T)$, but whose orientation coincides with the one of $\overline{\mathcal{A}}_{\mathbf{Y}}^d(x_0, x_1, \ldots, x_d, T)$ if and only if $r(T)$ is even.

Proposition A.25 *Let $T_0 \in \text{RTree}_d$ have a unique four-valent internal vertex while all other internal vertices are three-valent. Then the space $\mathcal{A}_{\mathbf{Y}}^d(x_0, x_1, \ldots, x_d, T_0)$ appears twice in the boundary of the oriented manifold*

$$\bigsqcup_{T \in \text{BinTree}_d} \widetilde{\mathcal{A}}_{\mathbf{Y}}^d(x_0, x_1, \ldots, x_d, T)$$

and the boundary orientations of the two copies are opposite to each other.

Proof Consider the union

$$\bigsqcup_{T \in \text{BinTree}_d} \overline{\mathcal{A}}_{\mathbf{Y}}^d(x_0, x_1, \ldots, x_d, T) \,.$$

There are two copies of $\mathcal{A}_{\mathbf{Y}}^d(x_0, x_1, \ldots, x_d, T_0)$ in the boundary of this union. By Lemma A.24, the boundary orientations of both of these copies have the same orientation. Introducing the additional orientation twists by $(-1)^{r(T)}$, Proposition A.19 implies that both copies of $\mathcal{A}_{\mathbf{Y}}^d(x_0, x_1, \ldots, x_d, T_0)$ will have opposite sign in $\bigsqcup_{T \in \text{BinTree}_d} \widetilde{\mathcal{A}}_{\mathbf{Y}}^d(x_0, x_1, \ldots, x_d, T)$. $\qquad\square$

By Proposition A.25, the orientation of $\bigsqcup_{T \in \mathrm{BinTree}_d} \widetilde{\mathcal{A}}_{\mathbf{Y}}^d(x_0, x_1, \ldots, x_d, T)$ induces an orientation of

$$\overline{\mathcal{A}}_{\mathbf{Y}}^d(x_0, x_1, \ldots, x_d) = \bigsqcup_{T \in \mathrm{BinTree}_d} \widetilde{\mathcal{A}}_{\mathbf{Y}}^d(x_0, x_1, \ldots, x_d, T) \Big/ \sim \,,$$

since the pairs of boundary curves that we are gluing have opposite orientations.

Throughout the rest of this section, let $\overline{\mathcal{A}}_{\mathbf{Y}}^d(x_0, x_1, \ldots, x_d)$ be equipped with the orientation that we have just described.

As we discussed in the proof of Corollary 6.27, the coefficients appearing in Theorem 6.28 are obtained by counting elements of moduli spaces which are components of the boundary of $\overline{\mathcal{A}}_{\mathbf{Y}}^d(x_0, x_1, \ldots, x_d)$. We have proven Corollary 6.27 by using the simple fact that the compact one-dimensional manifold $\overline{\mathcal{A}}_{\mathbf{Y}}^d(x_0, x_1, \ldots, x_d)$ has an even number of boundary points.

In an *oriented* compact one-dimensional manifold with boundary, every component with non-empty boundary is diffeomorphic to a compact interval. Hence, it has precisely two boundary points of which precisely one is positively oriented. If we denote the boundary orientation on $\partial \overline{\mathcal{A}}_{\mathbf{Y}}^d(x_0, x_1, \ldots, x_d)$ by

$$\epsilon_\partial : \partial \overline{\mathcal{A}}_{\mathbf{Y}}^d(x_0, x_1, \ldots, x_d) \to \{-1, 1\} \,,$$

this implies that

$$\sum_{(\gamma_1, \gamma_2) \in \partial \overline{\mathcal{A}}_{\mathbf{Y}}^d(x_0, x_1, \ldots, x_d)} \epsilon_\partial \left(\gamma_1, \gamma_2 \right) = 0 \,. \tag{A.14}$$

Equation (A.14) will be the decisive tool for the proof of Theorem 6.28. Using this equation, the proof reduces to a comparison of the boundary orientations and the oriented intersection numbers which define the coefficients appearing in the statement.

We will proceed along the following line of argument:

- We will compare the different orientations on the corresponding boundary components of the spaces $\mathcal{M}_{\mathbf{Y}}^d(x_0, x_1, \ldots, x_d, T)$, which are the domains of the endpoint evaluations used to define the intersection numbers. The three cases of convergence phenomena along the rooted edge, an internal edge and a leafed edge will be treated seperately, see Lemma A.27 and Theorem A.26.
- While these considerations suffice for external edges, we will compare the orientations of the different T-diagonals involved in the situation for internal edges, see Lemma A.28.
- Moreover, we compute the contribution of the twisting signs $(-1)^{\sigma(x_0, x_1, \ldots, x_d)}$ in the definition of the coefficients in Proposition A.31.
- Finally, we will conclude the proof by putting all the results together to compute the differences between the coefficients under investigation and the sums of the boundary orientations and use Eq. (A.14) to derive the statement.

For brevity's sake, we will denote the space of finite-length trajectories of positive length by

$$\mathcal{M} := \{(l, \gamma) \in \mathcal{M}(f, g) \mid l > 0\} .$$

We first deal with those components of $\partial \overline{\mathcal{A}}_{\mathbf{Y}}^d(x_0, x_1, \ldots, x_d)$ coming from the geometric convergence along an external edge.

Theorem A.26 *Let $T \in \mathrm{BinTree}_d$.*

(a) *Let $y_0 \in \mathrm{Crit} f$ satisfy $\mu(y_0) = \mu(x_0) - 1$. Then the product orientation on*

$$\widehat{\mathcal{M}}(x_0, y_0) \times \mathcal{A}_{\mathbf{Y}}^d(y_0, x_1, \ldots, x_d, T)$$

coincides with the boundary orientation with respect to $\overline{\mathcal{A}}_{\mathbf{Y}}^d(x_0, x_1, \ldots, x_d, T)$ if and only if $\mu(x_0)$ is odd.

(b) *For $i \in \{1, 2, \ldots, d\}$ let $y_i \in \mathrm{Crit} f$ with $\mu(y_i) = \mu(x_i) + 1$. Then the product orientation on*

$$\mathcal{A}_{\mathbf{Y}}^d(x_0, x_1, \ldots, x_{i-1}, y_i, x_{i+1}, \ldots, x_d, T) \times \widehat{\mathcal{M}}(y_i, x_i)$$

coincides with the induced boundary orientation with respect to $\overline{\mathcal{A}}_{\mathbf{Y}}^d(x_0, x_1, \ldots, x_d, T)$ if and only if the following number is even:

$$\mu(x_0) + 1 + (d - i)(n + 1) + \maltese_1^{i-1} .$$

Proof The main tool in proving the two parts of the theorem will be the construction of geometric gluing maps

$$G_0 : [\rho_0, +\infty) \times \widehat{\mathcal{M}}(x_0, y_0) \times \mathcal{A}_{\mathbf{Y}}^d(y_0, x_1, \ldots, x_d, T) \to \mathcal{A}_{\mathbf{Y}}^d(x_0, x_1, \ldots, x_d, T) , \qquad (A.15)$$

$$G_i : [\rho_i, +\infty) \times \mathcal{A}_{\mathbf{Y}}^d(x_0, x_1, \ldots, x_{i-1}, y_i, \ldots, x_d, T) \times \widehat{\mathcal{M}}(y_i, x_i) \to \mathcal{A}_{\mathbf{Y}}^d(x_0, x_1, \ldots, x_d, T) ,$$

for all $i \in \{1, 2, \ldots, d\}$. If we equip their domains with the product orientations, the definition of a geometric gluing map will imply that the G_i, where $i \in \{0, 1, \ldots, d\}$, will be orientation-preserving if and only if the product orientation of

$$\begin{cases} \widehat{\mathcal{M}}(x_0, y_0) \times \mathcal{A}_{\mathbf{Y}}^d(y_0, x_1, \ldots, x_d, T) & \text{for } i = 0, \\ \mathcal{A}_{\mathbf{Y}}^d(x_0, x_1, \ldots, x_{i-1}, y_i, x_{i+1}, \ldots, x_d, T) \times \widehat{\mathcal{M}}(y_i, x_i) & \text{for } i \in \{1, 2, \ldots, d\}, \end{cases}$$

coincides with its boundary orientation as a boundary of $\overline{\mathcal{A}}_{\mathbf{Y}}^d(x_0, x_1, \ldots, x_d, T)$.

(a) We first want to consider a geometric gluing map of the form

$$G_0' : [\rho_0, +\infty) \times \widehat{\mathcal{M}}(x_0, y_0) \times \mathcal{M}_{\mathbf{Y}}^d(y_0, x_1, \ldots, x_d, T) \to \mathcal{M}_{\mathbf{Y}}^d(x_0, x_1, \ldots, x_d, T) .$$

One checks that one obtains such a geometric gluing map by

$$G_0' := F_T^{-1} \circ G_0'' \circ (\mathrm{id}_{[\rho_0, +\infty) \times \widehat{\mathcal{M}}(x_0, y_0)} \times F_T') ,$$

where F_T and F_T' are the diffeomorphisms

$$F_T : \mathcal{A}_{\mathbf{Y}}^d(x_0, x_1, \ldots, x_d, T) \to \mathcal{W}^u(x_0) \times \prod_{e \in E_{int}(T)} \mathcal{M} \times \prod_{q=1}^d \mathcal{W}^s(x_q) ,$$

$$F_T' : \mathcal{A}_{\mathbf{Y}}^d(y_0, x_1, \ldots, x_d, T) \to \mathcal{W}^u(y_0) \times \prod_{e \in E_{int}(T)} \mathcal{M} \times \prod_{q=1}^d \mathcal{W}^s(x_q) ,$$

described in Sect. A.4, which are by definition orientation-preserving, and where

$$G_0'' : [\rho_0, +\infty) \times \widehat{\mathcal{M}}(x_0, y_0) \times \mathcal{W}^u(y_0) \times \prod_{e \in E_{int}(T)} \mathcal{M} \times \prod_{q=1}^d \mathcal{W}^s(x_q) \to$$

$$\mathcal{W}^u(x_0) \times \prod_{e \in E_{int}(T)} \mathcal{M} \times \prod_{q=1}^d \mathcal{W}^s(x_q) ,$$

where $\rho_0 > 0$ is sufficiently big, is defined as a composition of permutations of the factors and a gluing map for negative half-trajectories. We explicitly compute the sign of this map. We permute $[\rho_0, +\infty)$ along $\widehat{\mathcal{M}}(x_0, y_0) \times \mathcal{W}^u(y_0)$. The sign of this permutation is given by the parity of

$$\dim[\rho_0, +\infty) \cdot \left(\dim \widehat{\mathcal{M}}(x_0, y_0) + \dim \mathcal{W}^u(y_0)\right) = \mu(y_0) \equiv \mu(x_0) + 1 .$$

Afterwards, we apply a gluing map for negative half-trajectories

$$\widehat{\mathcal{M}}(x_0, y_0) \times \mathcal{W}^u(y_0) \times [\rho_0, +\infty) \to \mathcal{W}^u(x_0) ,$$

which is by definition orientation-preserving. Hence, the sign of G_0' is given by the parity of $\mu(x_0) + 1$. By standard gluing analysis methods of Morse theory, one shows that one can choose the gluing map in the construction of G_0'' in such a way that G_0' restricts to a map G_0 given as in (A.15) and the properties of G_0'' imply that G_0 is indeed a geometric gluing map. The domain and target of G_0 in (A.15) are obtained as the transverse intersections of the domain and target of G_0' with Δ_T. By Proposition A.23, the orientation of the space we obtain from intersecting $\widehat{\mathcal{M}}(x_0, y_0) \times \mathcal{M}_{\mathbf{Y}}^d(y_0, x_1, \ldots, x_d, T)$ with the boundary orientation with Δ_T differs from the boundary orientation of $\overline{\mathcal{A}}_{\mathbf{Y}}^d(x_0, x_1, \ldots, x_d, T)$ by the parity of

$$\operatorname{codim} \Delta_T = ((1 + 2k(T) + d) - (k(T) + 1))n = (k(T) + d)n = (2d - 2)n \equiv 0.$$

Thus, we have shown that the orientation of $\mathcal{A}_{\mathbf{Y}}^d(y_0, x_1, \ldots, x_d, T)$ coincides with the boundary orientation of $\overline{\mathcal{A}}_{\mathbf{Y}}^d(x_0, x_1, \ldots, x_d, T)$ if and only if $\mu(x_0) + 1$ is even.

(b) Fix $i \in \{1, 2, \ldots, d\}$. Similar to part (a), one constructs a geometric gluing map

$$G_i' : [\rho_0, +\infty) \times \mathcal{M}_{\mathbf{Y}}^d(y_0, x_1, \ldots, x_{i-1}, y_i, x_{i+1}, \ldots, x_d, T) \times \widehat{\mathcal{M}}(y_i, x_i)$$
$$\rightarrow \mathcal{M}_{\mathbf{Y}}^d(x_0, x_1, \ldots, x_d, T)$$

for sufficiently big $\rho_0 > 0$ that restricts to a gluing map of the form G_i, such that G_i' is orientation-preserving if and only if the following embedding is orientation-preserving, which is defined as a composition of permutations and a gluing map for positive half-trajectories:

$$[\rho_0, +\infty) \times \mathcal{W}^u(x_0) \times \prod_{e \in E_{int}(T)} \mathcal{M} \times \prod_{q=0}^{i-1} \mathcal{W}^s(x_q) \times \mathcal{W}^s(y_i) \times \prod_{q=i+1}^{d} \mathcal{W}^s(x_q)$$

$$\times \widehat{\mathcal{M}}(y_i, x_i) \rightarrow \mathcal{W}^u(x_0) \times \prod_{e \in E_{int}(T)} \mathcal{M} \times \prod_{q=1}^{d} \mathcal{W}^s(x_q).$$

We compute the sign of this map explicitly by decomposing it into three simpler steps.

Firstly, we permute the factor $\widehat{\mathcal{M}}(y_i, x_i)$ with $\prod_{q=i+1}^{d} \mathcal{W}^s(x_q)$. Since $\widehat{\mathcal{M}}(y_i, x_i)$ is zero-dimensional, this does not affect the sign. Secondly, we move $[\rho_0, +\infty)$ along $\mathcal{W}^u(x_0) \times \prod_{e \in E_{int}(T)} \mathcal{M} \times \prod_{q=1}^{i-1} \mathcal{W}^s(x_q) \times \mathcal{W}^s(y_i) \times \widehat{\mathcal{M}}(y_i, x_i)$. The sign of this permutation is given by the parity of

$$\dim \mathcal{W}^u(x_0) + \sum_{e \in E_{int}(T)} \dim \mathcal{M} + \sum_{q=1}^{i-1} \dim \mathcal{W}^s(x_q) + \dim \mathcal{W}^s(y_i) + \dim \widehat{\mathcal{M}}(y_i, x_i)$$

$$= \mu(x_0) + k(T)(n + 1) + \sum_{q=1}^{i-1}(n - \mu(x_q)) + n - \mu(y_i)$$

$$= \mu(x_0) + (d - 2)(n + 1) + in - \sum_{q=1}^{i-1} \mu(x_q) - \mu(y_i)$$

$$\equiv \mu(x_0) + (d - i)(n + 1) + \maltese_1^{i-1} + \mu(y_i) + 1.$$

Thirdly, we apply a gluing map $\mathcal{W}^s(y_i) \times \widehat{\mathcal{M}}(y_i, x_i) \times [\rho_0, +\infty) \rightarrow \mathcal{W}^s(x_i)$ for positive half-trajectories, whose sign is by Proposition A.6 given by the parity of $\mu(y_i)$. Hence, the total sign of the above diffeomorphism is given by the parity of

$$\mu(x_0) + 1 + (d - i)(n + 1) + \maltese_1^{i-1} \; .$$

Continuing the line of argument as in part 1 shows the claim.

□

We turn our attention to boundary spaces of $\overline{\mathcal{A}}_\mathbf{Y}^d(x_0, x_1, \ldots, x_d, T)$ that are obtained by geometric convergence along internal edges of T. The proof of the following lemma is given in strict analogy with the line of argument in [Abo09, Appendix C].

Lemma A.27 *Let* $T \in \mathrm{BinTree}_d$, $e \in E_{int}(T)$ *be of type* (i, l), $y \in \mathrm{Crit} f$ *satisfy*

$$\mu(y) = \sum_{q=i}^{i+l} \mu(x_q) + 1 - l \tag{A.16}$$

and let (T_1^e, T_2^e) *denote the splitting of* T *along* e. *For sufficiently big* $\rho_0 > 0$ *there exists a geometric gluing map*

$$G_{T,e} : [\rho_0, +\infty) \times \mathcal{M}_\mathbf{Y}^{d-l}(x_0, x_1, \ldots, x_{i-1}, y, x_{i+l+1}, \ldots, x_d, T_1^e) \times \mathcal{M}_\mathbf{Y}^{l+1}(y, x_i, \ldots, x_{i+l}, T_2^e)$$
$$\to \mathcal{M}_\mathbf{Y}^d(x_0, x_1, \ldots, x_d, T) \, ,$$

which is orientation-preserving if and only if the following number is even:

$$\begin{cases} \mu(x_0) + \maltese_1^{i-1} + (n + 1)\left(l \cdot \maltese_1^{i-1} + dl + i + l + d + 1\right) & \text{if } e \text{ is left-handed}, \\ \mu(x_0) + \maltese_1^{i-1} + (n + 1)\left(l \cdot \maltese_1^{i-1} + dl + i + l + d\right) & \text{if } e \text{ is right-handed}. \end{cases}$$

Proof We consider a map

$$G'_{T,e} : [\rho_0, \infty) \times \mathcal{W}^u(x_0) \times \prod_{e' \in E_{int}(T_1^e)} \mathcal{M} \times \prod_{q=1}^{i-1} \mathcal{W}^s(x_q) \times \mathcal{W}^s(y) \times \prod_{q=i+l+1}^{d} \mathcal{W}^s(x_q)$$

$$\times \, \mathcal{W}^u(y) \times \prod_{e' \in E_{int}(T_2^e)} \mathcal{M} \times \prod_{q=i}^{i+l} \mathcal{W}^s(x_q) \to \mathcal{W}^u(x_0) \times \prod_{e' \in E_{int}(T)} \mathcal{M} \times \prod_{q=1}^{d} \mathcal{W}^s(x_q)$$

which is defined in the obvious way as a composition of permutations and a gluing map for finite-length trajectories. Via the maps $F_{T_1^e}$, $F_{T_2^e}$ and F_T from Sect. A.4, domain and target of $G'_{T,e}$ are identified with the domain and target of the map $G_{T,e}$ that we want to show to exist. Indeed, one obtains a geometric gluing map of the form $G_{T,e}$ by composing $G'_{T,e}$ with $F_{T_1^e} \times F_{T_2^e}$ and F_T^{-1} in analogy with the line of argument in the proof of Theorem A.26. Moreover, $G_{T,e}$ is orientation-preserving if and only if $G'_{T,e}$ is and we will compute the sign of $G'_{T,e}$ explicitly. We start by computing the sign of the permutation which maps the domain of $G'_{T,e}$ onto

$$\mathcal{W}^u(x_0) \times [\rho_0, \infty) \times \mathcal{W}^s(y) \times \mathcal{W}^u(y) \times \prod_{e' \in E_{int}(T_1^e)} \mathcal{M} \times \prod_{e' \in E_{int}(T_2^e)} \mathcal{M} \times \prod_{q=1}^{d} \mathcal{W}^s(x_q) \,.$$

(A.17)

We view this permutation as a composition of simpler permutations in the following way:

1. We move $\mathcal{W}^s(y)$ past the factors $\mathcal{W}^s(x_1) \times \cdots \times \mathcal{W}^s(x_{i-1})$. The sign of this permutation is given by the parity of

$$\dim \mathcal{W}^s(y) \cdot \sum_{q=1}^{i-1} \dim \mathcal{W}^s(x_q) = (n - \mu(y))\Big((i-1)n - \sum_{q=1}^{i-1} \mu(x_q)\Big)$$

$$\equiv n\Big((i-1)\mu(y) + \maltese_1^{i-1}\Big) + \mu(y)\sum_{q=1}^{i-1} \mu(x_q) \,. \quad (A.18)$$

2. In the modified product, we move $\mathcal{W}^u(y)$ past

$$\mathcal{W}^s(x_1) \times \cdots \times \mathcal{W}^s(x_{i-1}) \times \mathcal{W}^s(x_{i+l+1}) \times \cdots \times \mathcal{W}^s(x_d) \,.$$

The sign of this permutation is given by the parity of

$$\dim \mathcal{W}^u(y) \cdot \Big(\sum_{q=1}^{i-1} \dim \mathcal{W}^s(x_q) + \sum_{q=i+l+1}^{d} \dim \mathcal{W}^s(x_q)\Big)$$

$$= \mu(y)\Big((d-l-1)n - \sum_{q=1}^{i-1} \mu(x_q) - \sum_{q=i+l+1}^{d} \mu(x_q)\Big) \,. \quad (A.19)$$

3. Afterwards, we move $\prod_{e' \in E_{int}(T_2^e)} \mathcal{M}$ past

$$\mathcal{W}^s(y) \times \mathcal{W}^u(y) \times \mathcal{W}^s(x_1) \times \cdots \times \mathcal{W}^s(x_{i-1}) \times \mathcal{W}^s(x_{i+l+1}) \times \cdots \times \mathcal{W}^s(x_d) \,.$$

This affects the sign by the parity of

$|E_{int}(T_2^e)| \cdot \dim \mathcal{M} \cdot$

$$\left(\dim \mathcal{W}^s(y) + \dim \mathcal{W}^u(y) + \sum_{q=1}^{i-1} \dim \mathcal{W}^s(x_q) + \sum_{q=i+l+1}^{d} \dim \dot{\mathcal{W}}^s(x_q) \right)$$

$$= (l-1)(n+1)\left(n + \sum_{q=1}^{i-1}(n-\mu(x_q)) + \sum_{q=i+l+1}^{d}(n-\mu(x_q)) \right)$$

$$= (l-1)(n+1)\left((d-l)n - \sum_{q=1}^{i-1}\mu(x_q) - \sum_{q=i+l+1}^{d}\mu(x_q) \right)$$

$$\equiv (l-1)(n+1)\left(\sum_{q=1}^{i-1}\mu(x_q) + \sum_{q=i+l+1}^{d}\mu(x_q) \right). \tag{A.20}$$

4. We permute $\mathcal{W}^s(x_i) \times \cdots \times \mathcal{W}^s(x_{i+l})$ past $\mathcal{W}^s(x_{i+l+1}) \times \cdots \times \mathcal{W}^s(x_d)$ such that all the $\mathcal{W}^s(x_q)$ appear in the right order. The sign of this permutation is given by the parity of

$$\sum_{q=i}^{i+l} \dim \mathcal{W}^s(x_q) \cdot \sum_{q=i+l+1}^{d} \dim \mathcal{W}^s(x_q) = \sum_{q=i}^{i+l}(n-\mu(x_q)) \cdot \sum_{q=i+l+1}^{d}(n-\mu(x_q))$$

$$= \left((l+1)n - \sum_{q=i}^{i+l}\mu(x_q) \right) \cdot \left((d-i-l)n - \sum_{q=i+l+1}^{d}\mu(x_q) \right)$$

$$\equiv (d-i-l)(l+1)n + n\left((d-i-l)\sum_{q=i}^{i+l}\mu(x_q) + (l+1)\sum_{q=i+l+1}^{d}\mu(x_q) \right)$$

$$+ \sum_{q=i}^{i+l}\mu(x_q) \cdot \sum_{q=i+l+1}^{d}\mu(x_q)$$

$$= n\left((d-i-l)\left(\sum_{q=i}^{i+l}\mu(x_q) + l + 1 \right) + (l+1)\sum_{q=i+l+1}^{d}\mu(x_q) \right)$$

$$+ \sum_{q=i}^{i+l}\mu(x_q) \cdot \sum_{q=i+l+1}^{d}\mu(x_q). \tag{A.21}$$

5. We permute $\mathcal{W}^s(y) \times \mathcal{W}^u(y)$ along $\prod_{e' \in E_{int}(T_1^e)} \mathcal{M} \times \prod_{e' \in E_{int}(T_2^e)} \mathcal{M}$. The sign of this permutation is given by

$$(\dim \mathcal{W}^s(y) + \dim \mathcal{W}^u(y)) \cdot \left(\sum_{e' \in E_{int}(T_1^e) \cup E_{int}(T_2^e)} \dim \mathcal{M} \right) \equiv n \cdot (n+1) \cdot (d-1) \equiv 0.$$

6. Finally, we permute $[\rho_0, \infty)$ past $\mathcal{W}^u(x_0)$, which modifies the sign by the parity of

$$\dim \mathcal{W}^u(x_0) = \mu(x_0) . \tag{A.22}$$

The sign of the permutation under investigation is given by the parity of the sum of (A.18)–(A.22). The parity of the sum of (A.18) and (A.19) is given by

$$n\left((d - l - i)\mu(y) + \maltese_1^{i-1}\right) + \mu(y) \sum_{q=i+l+1}^{d} \mu(x_q) .$$

Adding (A.21) to this result and taking the congruence class modulo two yields

$$n\left((d - l - i)\left(\mu(y) - \sum_{q=i}^{i+l} \mu(x_q) + l - 1\right) + \maltese_1^{i-1}\right)$$

$$+ \left(\mu(y) - \sum_{q=i}^{i+l} \mu(x_q) + l - 1 + (l-1)(n+1)\right) \cdot \sum_{q=i+l+1}^{d} \mu(x_q)$$

$$\equiv n \cdot \maltese_1^{i-1} + (l-1)(n+1) \cdot \sum_{q=i+l+1}^{d} \mu(x_q) ,$$

where we used (A.16) for the last congruence. Adding (A.20) and considering again congruence modulo two yields

$$n \cdot \maltese_1^{i-1} + (l-1)(n+1) \cdot \sum_{q=1}^{i-1} \mu(x_q) \equiv n \cdot \maltese_1^{i-1} + (l-1)(n+1)\left(\maltese_1^{i-1} + i - 1\right)$$

$$\equiv \maltese_1^{i-1} + (n+1)\left(l \cdot \maltese_1^{i-1} + (i-1)(l-1)\right) .$$

If we finally add (A.22), we obtain

$$\mu(x_0) + \maltese_1^{i-1} + (n+1)\left(l \cdot \maltese_1^{i-1} + (i-1)(l-1)\right)$$

$$\equiv \mu(x_0) + \maltese_1^{i-1} + (n+1)\left(l \cdot \maltese_1^{i-1} + il + i + l + 1\right) .$$

This is the parity of the sign of the permutation which permutes the domain of our map onto (A.17). Since the gluing map

$$[\rho_0, \infty) \times \mathcal{W}^s(y) \times \mathcal{W}^u(y) \to \mathcal{M}(f, g)$$

has sign $(-1)^{n+1}$ by Proposition A.8, we need to add $n + 1$ to the above, which yields:

$$\mu(x_0) + \maltese_1^{i-1} + (n+1)\left(l \cdot \maltese_1^{i-1} + il + i + l\right) . \tag{A.23}$$

The final step is to take into account the sign of the permutation

$$\mathcal{M} \times \prod_{e' \in T_1^e} \mathcal{M} \times \prod_{e' \in T_2^e} \mathcal{M} \to \prod_{e' \in E_{int}(T)} \mathcal{M} \qquad (A.24)$$

according to the ordering of the edges of the trees. This sign is obviously given by

$$(-1)^{n+1} \cdot \text{sign } \tau_e ,$$

where τ_e again denotes the order-preserving permutation

$$\tau_e : \{e\} \times E_{int}(T_1^e) \times E_{int}(T_2^e) \to E_{int}(T) .$$

Since e is of type (i, l), it follows from Theorem A.16 that the sign of (A.24) is given by the parity of

$$\begin{cases} (n + 1)((d - i)l + d - 1) & \text{if } e \text{ is left-handed,} \\ (n + 1)((d - i)l + d) & \text{if } e \text{ is right-handed.} \end{cases}$$

Adding this to (A.23) shows the claim.   □

The definition of the oriented intersection numbers $\#_{or}\mathcal{A}_Y^d(x_0, x_1, \ldots, x_d, T)$ relies on the orientations of the T-diagonals. Therefore, we need to establish the following lemma to compare the boundary orientations under consideration:

Lemma A.28 *Let $T \in \text{RTree}_d$ and let $e \in E_{int}(T)$ be of type (i, l). The diffeomorphism*

$$s_{T,e} : \Delta_{T_1^e} \times \Delta_{T_2^e} \to \Delta_T , \qquad (A.25)$$

given as in Lemma 6.13, is orientation-preserving if and only if the following number is even:

$$\begin{cases} n((d - i - 1)l + 1) & \text{if } e \text{ is left-handed,} \\ n(d - i - 1)l & \text{if } e \text{ is right-handed.} \end{cases}$$

It extends to a diffeomorphism $\sigma_{T,e} : M^{1+2k(T_1^e)+d-l} \times M^{1+2k(T_2^e)+l+1} \to M^{1+2k(T)+d}$ *which is orientation-preserving if and only if $n(d - i - 1)l$ is even.*

Proof One checks in the proof of Lemma 6.13 that the diffeomorphism from (A.25) is given as the restriction of a diffeomorphism

$$\sigma_{T,e} : M^{1+2k(T_1^e)+d-l} \times M^{1+2k(T_2^e)+l+1} \to M^{1+2k(T)+d}$$

given by permuting factors of the products, where we put $k_1 := k(T_1^e)$ and $k_2 := k(T_2^e)$. We first compute the sign of $\sigma_{T,e}$ by decomposing it into a sequence of simpler permutations and compute the respective signs:

1. We permute the ith of the last $d - l$ factors of M^{1+2k_1+d-l} along M^{2k_1+i-1}. The sign of this map is given by the parity of

$$(2k_1 + i - 1)n \equiv (i - 1)n . \tag{A.26}$$

2. We move the first factor of M^{1+2k_2+l+1} along $M^{2k_1+d-l-1}$. This affects the sign by the parity of

$$(2k_1 + d - l - 1)n \equiv (d - l - 1)n . \tag{A.27}$$

3. We interchange M^{2k_2} with M^{d-l-1}. This permutation is orientation-preserving, since M^{2k_2} is always even-dimensional.

4. We move M^{l+1} along M^{d-i-l}. The sign of this map is given by the parity of

$$n(d - i - l)(l + 1) \equiv n(d - i)(l + 1) . \tag{A.28}$$

5. Finally, we need to permute $M \times M^2 \times M^{2k_1} \times M^{2k_2} \times M^d$ onto $M \times M^{2k} \times M^d$ according to the orderings of $E_{int}(T_1^e)$, $E_{int}(T_2^e)$ and $E_{int}(T)$. This permutation is always orientation-preserving, because we are only moving even-dimensional manifolds.

Therefore, the total sign of the permutation $M^{1+2k_1+d-l} \times M^{1+2k_2+l+1} \xrightarrow{\cong} M^{1+2k+d}$ is given by the parity of the sum of (A.26)–(A.28), which amounts to

$$n((d - i)(l + 1) + d - i - l) \equiv n((d - i)l - l) \equiv n(d - i - 1)l .$$

We next compute the sign of the map $s_{T,e} := \sigma_{T,e}|_{\Delta_{T_1^e} \times \Delta_{T_2^e}} : \Delta_{T_1^e} \times \Delta_{T_2^e} \to \Delta_{T,}$. We have defined the orientations on $\Delta_{T_1^e}$, $\Delta_{T_2^e}$ and Δ_T such that there are orientation-preserving diffeomorphisms $M^{1+k_1} \xrightarrow{\cong} \Delta_{T_1^e}$, $M^{1+k_2} \xrightarrow{\cong} \Delta_{T_2^e}$ and $M^{1+k} \xrightarrow{\cong} \Delta_T$.

Consequently, the sign of $s_{T,e}$ is given by the sign of the diffeomorphism

$$M^{1+k_1} \times M^{1+k_2} \to M^{1+k} ,$$

obtained via moving the first factor of M^{1+k_2} along the last k_1 of the product M^{1+k_1} and afterwards permute the last $1 + k_1 + k_2$ factors of $M^{1+1+k_1+k_2}$ according to the ordering of $E_{int}(T)$. Since T_1^e is binary, it holds that $k_1 = d - l - 2$. Consequently, the sign is given by the parity of

$$n(d - l - 2 + \mathrm{sign}\tau_e) \equiv n(d - l + \mathrm{sign}\tau_e) \equiv \begin{cases} n((d - i - 1)l + 1) & \text{if } e \text{ is left-handed,} \\ n(d - i - 1)l & \text{if } e \text{ is right-handed,} \end{cases}$$

where we have uses Theorem A.16. This completes the proof. $\qquad\square$

The following lemma will turn out to be useful in several of the upcoming sign computations.

Lemma A.29 *Let $r \in \mathbb{N}$, $T \in \mathrm{RTree}_r$ and let $y_0, y_1, \ldots, y_r \in \mathrm{Crit} f$ satisfy*

$$\mu(y_0) = \sum_{q=1}^{r} \mu(y_q) + 2 - r .$$

If T is a binary tree, then $\mathcal{M}_{\mathbf{Y}}^r(y_0, y_1, \ldots, y_r, T)$ will be even-dimensional.

Proof If T is binary, then $k(T) = r - 2$ and we compute that

$$\dim \mathcal{M}_{\mathbf{Y}}^r(y_0, y_1, \ldots, y_r, T) = \mu(y_0) + k(T)(n + 1) + \sum_{q=1}^{r} (n - \mu(y_q))$$

$$= \mu(y_0) - \sum_{q=1}^{r} \mu(y_q) + (r - 2)(n + 1) + rn = 2 - r + (2r - 2)n + r - 2 = 2(r - 1)n .$$

$\square$

Theorem A.30 *Let $T \in \mathrm{BinTree}_d$, $e \in E_{int}(T)$ be of type (i, l), $y \in \mathrm{Crit} f$ satisfy*

$$\mu(y) = \sum_{q=i}^{i+l} \mu(x_q) + 1 - l .$$

and let (T_1^e, T_2^e) be the splitting of T along e. The product orientation of

$$\mathcal{A}_{\mathbf{Y}}^{d-l}(x_0, x_1, \ldots, x_{i-1}, y, x_{i+l+1}, \ldots, x_d, T_1^e) \times \mathcal{A}_{\mathbf{Y}}^{l+1}(y, x_i, \ldots, x_{i+l}, T_2^e)$$

coincides with its boundary orientation with respect to $\overline{\mathcal{A}}_{\mathbf{Y}}^d(x_0, x_1, \ldots, x_d, T)$ if and only if the following number is even:

$$r(T_1^e) + r(T_2^e) + r(T) + \mu(x_0) + 1 + \maltese_1^{i-1} + (n + 1)\left(l \cdot \maltese_1^{i-1} + dl + i + l + d\right) .$$

Proof By standard methods of gluing analysis in Morse theory, one shows that there exists a geometric gluing map of the form $G_{T,e}$ as in Lemma A.27, which restricts to a map

$$G : [\rho_0, +\infty) \times \mathcal{A}_{\mathbf{Y}}^{d-l}(x_0, x_1, \ldots, x_{i-1}, y, x_{i+l+1}, \ldots, x_d, T_1^e) \times \mathcal{A}_{\mathbf{Y}}^{l+1}(y, x_i, \ldots, x_{i+l}, T_2^e)$$

$$\to \mathcal{A}_{\mathbf{Y}}^d(x_0, x_1, \ldots, x_d, T) .$$

In analogy with the line of argument in the proof of Theorem A.26 the two orientations under consideration coincide if and only if G is orientation-preserving. In the following, we will compute the sign of G explicitly. The orientation of the domain of G is given by the product of

- the standard orientation of $[\rho_0, +\infty)$,

- the orientation induced by the transverse intersection of

$$\mathcal{M}_{\mathbf{Y}}^{d-l}(x_0, x_1, \ldots, x_{i-1}, y, x_{i+l+1}, \ldots, x_d, T_1^e)$$

with $\Delta_{T_1^e}$ under $\underline{E}_{\mathbf{Y}}$,
- the orientation induced by the transverse intersection of $\mathcal{M}_{\mathbf{Y}}^{l+1}(y, x_i, \ldots, x_{i+l}, T_2^e)$ with $\Delta_{T_2^e}$ under $\underline{E}_{\mathbf{Y}}$.

We want to compare this orientation to the one induced by the transverse intersection of

$$[\rho_0, +\infty) \times \mathcal{M}_{\mathbf{Y}}^{d-l}(x_0, x_1, \ldots, x_{i-1}, y, \ldots, x_d, T_1^e) \times \mathcal{M}_{\mathbf{Y}}^{l+1}(y, x_i, \ldots, x_{i+l}, T_2^e) \tag{A.29}$$

with $\Delta_{T_1^e} \times \Delta_{T_2^e}$ under the product of the evaluation maps. In the zero-dimensional case which we are considering, these two orientations coincide if and only if the following number is even:

$$\dim \mathcal{M}_{\mathbf{Y}}^{l+1}(y, x_i, \ldots, x_{i+l}, T_2^e) \cdot \dim \Delta_{T_1^e} \equiv 0 ,$$

where we have applied Lemma A.29, using that T_2^e is a binary tree. Thus, the two orientations coincide such that it suffices to compare the orientation induced by the transverse intersection of (A.29) with $\Delta_{T_1^e} \times \Delta_{T_2^e}$ with the orientation of $\mathcal{A}_{\mathbf{Y}}^d(x_0, x_1, \ldots, x_d)$. This situation is a special case of Theorem A.10. More precisely, that theorem applies with

$$M_1 = [\rho_0, +\infty) \times \mathcal{M}_{\mathbf{Y}}^{d-l}(x_0, x_1, \ldots, x_{i-1}, y, x_{i+l+1}, \ldots, x_d, T_1^e) \times \mathcal{M}_{\mathbf{Y}}^{l+1}(y, x_i, \ldots, x_{i+l}, T_2^e) ,$$

$$S_1 = [\rho_0, +\infty) \times \mathcal{A}_{\mathbf{Y}}^{d-l}(x_0, x_1, \ldots, x_{i-1}, y, x_{i+l+1}, \ldots, x_d, T_1^e) \times \mathcal{A}_{\mathbf{Y}}^{l+1}(y, x_i, \ldots, x_{i+l}, T_2^e) ,$$

$$M_2 = \mathcal{M}_{\mathbf{Y}}^d(x_0, x_1, \ldots, x_d, T) , \qquad S_2 = \mathcal{A}_{\mathbf{Y}}^d(x_0, x_1, \ldots, x_d, T) ,$$

$$N_1 = \Delta_{T_1^e} \times \Delta_{T_2^e} , \qquad N_2 = \Delta_T ,$$

$$\varphi = G_{T,e} , \qquad \psi = \sigma_{T,e} , \qquad f_1 = \underline{E}_{\mathbf{Y}} \times \underline{E}_{\mathbf{Y}} , \qquad f_2 = \underline{E}_{\mathbf{Y}} ,$$

where $G_{T,e}$ is suitably chosen as explained above and where $\sigma_{T,e}$ is the diffeomorphism from Lemma A.28. If e is left-handed, then combining Theorem A.10 with Lemmas A.27 and A.28 will yield that $G_{T,e}|_{S_1}$ is orientation-preserving if and only if the following number is even:

$$\mu(x_0) + \maltese_1^{i-1} + (n+1)\left(l \cdot \maltese_1^{i-1} + dl + i + l + d + 1\right)$$
$$+ n((d-i-1)l - 1) + n(d-i-1)l$$
$$\equiv \mu(x_0) + \maltese_1^{i-1} + (n+1)\left(l \cdot \maltese_1^{i-1} + dl + i + l + d + 1\right) + n$$
$$\equiv \mu(x_0) + 1 + \maltese_1^{i-1} + (n+1)\left(l \cdot \maltese_1^{i-1} + dl + i + l + d\right) .$$

If e is right-handed, then one shows along the same lines that the sign of $G_{T,e}|_{S_1}$ is given by the parity of

$$\mu(x_0) + \maltese_1^{i-1} + (n+1)\left(l \cdot \maltese_1^{i-1} + dl + i + l + d\right) .$$

The claim follows from applying part (b) of Proposition A.19.   □

The last ingredient for the proof of Theorem 6.28 is the computation of the con-tribution of the signs $(-1)^{\sigma(x_0, x_1, \ldots, x_d)}$. The occurring cases are subsumed in Propo-sition A.31. We will eventually see in the proof of Theorem 6.28 that the twisting numbers $\sigma(x_0, x_1, \ldots, x_d)$ are defined such that their contributions ensure the valid-ity of the desired equations.

Proposition A.31 Let $x_0, x_1, \ldots, x_d \in \operatorname{Crit} f$ satisfy $\mu(x_0) = \sum_{q=1}^{d} \mu(x_q) + 3 - d$.

1. Let $i \in \{1, 2, \ldots, d\}$, $l \in \{0, 1, \ldots, d-i\}$ and $y \in \operatorname{Crit} f$ satisfying

$$\mu(y) = \sum_{q=i}^{i+l} \mu(x_q) + 1 - l . \tag{A.30}$$

Then the following congruence holds modulo two:

$$\sigma(x_0, x_1, \ldots, x_d) + \sigma(x_0, x_1, \ldots, x_{i-1}, y, x_{i+l+1}, \ldots, x_d) + \sigma(y, x_i, \ldots, x_{i+l})$$
$$\equiv (n+1)\left(l \cdot \maltese_1^{i-1} + dl + d + l + i\right) .$$

2. Let $y_0 \in \operatorname{Crit} f$ with $\mu(y_0) = \mu(x_0) - 1$. The following congruence holds modulo two:
$$\sigma(x_0, y_0) + \sigma(y_0, x_1, \ldots, x_d) \equiv \sigma(x_0, x_1, \ldots, x_d) .$$

3. For $i \in \{1, 2, \ldots, d\}$ and $y_i \in \operatorname{Crit} f$ with $\mu(y_i) = \mu(x_i) + 1$ the following con-gruence holds modulo two:

$$\sigma(x_0, x_1, \ldots, x_{i-1}, y_i, x_{i+1}, \ldots, x_d) + \sigma(y_i, x_i) + \sigma(x_0, x_1, \ldots, x_d) \equiv (n+1)(d-i) .$$

Proof We will show all parts of the lemma by explicit and elementary computations.

1. We explicitly compute

$$\sigma(x_0, x_0, x_1, \ldots, x_{i-1}, y, x_{i+l+1}, \ldots, x_d)$$
$$= (n+1)\Big(\mu(x_0) + \sum_{j=1}^{i-1}(d - l + 1 - j)\mu(x_j) + (d - l + 1 - i)\mu(y)$$
$$+ \sum_{k=i+1}^{d-l}(d - l + 1 - k)\mu(x_{k+l})\Big)$$

$$= (n+1)\Big(\mu(x_0) + \sum_{j=1}^{i-1}(d-l+1-j)\mu(x_j) + (d-l+1-i)\mu(y)$$

$$+ \sum_{k=i+l+1}^{d}(d+1-k)\mu(x_k)\Big),$$

$$\sigma(y, x_i, \ldots, x_{i+l}) = (n+1)\Big(\mu(y) + \sum_{k=1}^{l+1}(l+2-k)\mu(x_{i-1+k})\Big)$$

$$= (n+1)\Big(\mu(y) + \sum_{k=i}^{i+l}(l+1-k-i)\mu(x_k)\Big).$$

Consequently,

$$\sigma(x_0, x_0, x_1, \ldots, x_{i-1}, y, x_{i+l+1}, \ldots, x_d) + \sigma(y, x_i, \ldots, x_{i+l})$$

$$\equiv (n+1)\Big(\mu(x_0) + \sum_{j=1}^{i-1}(d-l+1-j)\mu(x_j) + \sum_{k=i}^{i+l}(l+1-k-i)\mu(x_k)$$

$$+ \sum_{k=i+l+1}^{d}(d+1-k)\mu(x_k) + (d-l-i)\mu(y)\Big).$$

By definition of $\sigma(x_0, x_1, \ldots, x_d)$, this implies

$$\sigma(x_0, x_1, \ldots, x_d) + \sigma(x_0, x_0, x_1, \ldots, x_{i-1}, y, x_{i+l+1}, \ldots, x_d) + \sigma(y, x_i, \ldots, x_{i+l})$$

$$\equiv (n+1)\Big(2\mu(x_0) + \sum_{j=1}^{i-1}((d+1-j) + (d+1-l-j))\mu(x_j)$$

$$+ \sum_{k=i}^{i+l}((d+1-k) + (l+1-k-i))\mu(x_k) + (d-i-l)\mu(y)\Big)$$

$$\equiv (n+1)\Big(l \cdot \sum_{j=1}^{i-1}\mu(x_j) + (d-i-l)\Big(\mu(y) - \sum_{k=i}^{i+l}\mu(x_k)\Big)\Big)$$

$$\stackrel{(A.30)}{\equiv} (n+1)\Big(l \cdot \sum_{j=1}^{i-1}\mu(x_j) + (d-i)(l-1)\Big) \equiv (n+1)\Big(l \cdot \maltese_1^{i-1} + dl + d + l + i\Big).$$

2. We compute that $\sigma(x_0, y_0) = (n+1)(\mu(x_0) - \mu(y_0)) = n+1$. Moreover,

$$\sigma(y_0, x_1, \ldots, x_d) = (n+1)\Big(\mu(x_0) - 1 - \sum_{i=1}^{d}(d+1-i)\mu(x_i)\Big)$$

$$\equiv n+1 + \sigma(x_0, x_1, \ldots, x_d) = \sigma(x_0, y_0) + \sigma(y_0, x_1, \ldots, x_d).$$

The claim immediately follows.

3. We compute that $\sigma(y_i, x_i) = (n+1)(\mu(y_i) - \mu(x_i)) = n+1$. Moreover,

$$\sigma(x_0, x_1, \ldots, x_{i-1}, y_i, x_{i+1}, \ldots, x_d)$$

$$= (n+1) \cdot$$

$$\left(\mu(x_0) - \sum_{q=1}^{i-1}(d+1-q)\mu(x_q) - (d+1-i)(\mu(x_i)+1)\right.$$

$$\left. - \sum_{q=i+1}^{d}(d+1-q)\mu(x_q)\right)$$

$$\equiv \sigma(x_0, x_1, \ldots, x_d) + (n+1)(d+1-i)$$

$$\equiv \sigma(x_0, x_1, \ldots, x_d) + \sigma(y_i, x_i) + (n+1)(d-i) .$$

Again, the claim immediately follows.

$\square$

We have collected all ingredients required to prove Theorem 6.28.

Proof of Theorem 6.28 By definition of the coefficients, it holds for all $i \in \{1, \ldots, d-1\}$, $l \in \{1, 2, \ldots, d-1-i\}$ and $y \in \text{Crit} f$ of the right index given that

$$a_{\mathbf{Y}}^{d-l}(x_0, x_1, \ldots, x_{i-1}, y, x_{i+l+1}, \ldots, x_d) \cdot a_{\mathbf{Y}}^{l+1}(y, x_i, \ldots, x_{i+l})$$

$$= \sum_{(T_1, T_2)} (-1)^{\sigma(x_0, x_1, \ldots, x_{i-1}, y, x_{i+l+1}, \ldots, x_d) + r(T_1) + \sigma(y, x_i, \ldots, x_{i+l}) + r(T_2)}$$

$$\#_{\text{or}} \mathcal{A}_{\mathbf{Y}}^{d-l}(x_0, x_1, \ldots, x_{i-1}, y, x_{i+l+1}, \ldots, x_d, T_1) \cdot \#_{\text{or}} \mathcal{A}_{\mathbf{Y}}^{l+1}(y, x_i, \ldots, x_{i+l}, T_2)$$

$$= \sum_{T \in \text{BinTree}_d} \sum_{e \text{ of type } (i,l)} (-1)^{\sigma(x_0, x_1, \ldots, x_{i-1}, y, x_{i+l+1}, \ldots, x_d) + \sigma(y, x_i, \ldots, x_{i+l}) + r(T_1^e) + r(T_2^e)}$$

$$\#_{\text{or}} \mathcal{A}_{\mathbf{Y}}^{d-l}(x_0, x_1, \ldots, x_{i-1}, y, x_{i+l+1}, \ldots, x_d, T_1^e) \cdot \#_{\text{or}} \mathcal{A}_{\mathbf{Y}}^{l+1}(y, x_i, \ldots, x_{i+l}, T_2^e) ,$$

where we used Lemma 6.23 for the last equality. Explicitly, this number reads as

$$a_{\mathbf{Y}}^{d-l}(x_0, x_1, \ldots, x_{i-1}, y, x_{i+l+1}, \ldots, x_d) \cdot a_{\mathbf{Y}}^{l+1}(y, x_i, \ldots, x_{i+l}) \tag{A.31}$$

$$= \sum_{T \in \text{BinTree}_d} \sum_{e} \sum_{(\gamma_1, \gamma_2)} (-1)^{\sigma(x_0, x_1, \ldots, x_{i-1}, y, \ldots, x_d) + \sigma(y, x_i, \ldots, x_{i+l}) + r(T_1^e) + r(T_2^e)} \epsilon_{T_1^e}(\gamma_1) \cdot \epsilon_{T_2^e}(\gamma_2) ,$$

where the sum over $\left(\gamma_1, \gamma_2\right)$ in the last line is taken over all

$$\left(\gamma_1, \gamma_2\right) \in \mathcal{A}_{\mathbf{Y}}^{d-l}(x_0, x_1, \ldots, x_{i-1}, y, x_{i+l+1}, \ldots, x_d, T_1^e) \times \mathcal{A}_{\mathbf{Y}}^{l+1}(y, x_i, \ldots, x_{i+l}, T_2^e) =: \mathcal{A}_e .$$

As a consequence of Theorem A.30,

$$(-1)^{r(T_1^e) + r(T_2^e)} \epsilon_{T_1^e}(\gamma_1) \cdot \epsilon_{T_2^e}(\gamma_2)$$

$$= (-1)^{r(T) + \mu(x_0) + 1 + \boxtimes_1^{i-1} + (n+1)(l \cdot \boxtimes_1^{i-1} + dl + i + l + d)} \epsilon_{\partial, T}\left(\gamma_1, \gamma_2\right) \quad \forall \left(\gamma_1, \gamma_2\right) \in \mathcal{A}_e ,$$

where $\epsilon_{\partial,T}$ denotes the boundary orientation of $\overline{\mathcal{A}}_{\mathbf{Y}}^d(x_0, x_1, \ldots, x_d, T)$. By definition of the orientation on $\overline{\mathcal{A}}_{\mathbf{Y}}^d(x_0, x_1, \ldots, x_d)$, it holds that

$$\epsilon_\partial\left(\underline{\gamma}_1, \underline{\gamma}_2\right) = (-1)^{r(T)}\epsilon_{\partial,T}\left(\underline{\gamma}_1, \underline{\gamma}_2\right) \quad \forall \left(\underline{\gamma}_1, \underline{\gamma}_2\right) \in \mathcal{A}_e \,,$$

where $\epsilon_\partial : \partial\overline{\mathcal{A}}_{\mathbf{Y}}^d(x_0, x_1, \ldots, x_d) \to \{-1, 1\}$ denotes the boundary orientation of the glued manifold $\overline{\mathcal{A}}_{\mathbf{Y}}^d(x_0, x_1, \ldots, x_d)$. Thus,

$$(-1)^{r(T_1^e)+r(T_2^e)}\epsilon_{T_1^e}\left(\underline{\gamma}_1\right) \cdot \epsilon_{T_2^e}\left(\underline{\gamma}_2\right) = (-1)^{\mu(x_0)+1+\maltese_1^{i-1}+(n+1)\left(l\cdot\maltese_1^{i-1}+dl+i+l+d\right)}\epsilon_\partial\left(\underline{\gamma}_1, \underline{\gamma}_2\right)$$

and part 1 of Proposition A.31 implies

$$(-1)^{\sigma(x_0,x_1,\ldots,x_{i-1},y,\ldots,x_d)+r(T_1^e)+\sigma(y,x_i,\ldots,x_{i+l})+r(T_2^e)}\epsilon_{T_1^e}\left(\underline{\gamma}_1\right) \cdot \epsilon_{T_2^e}\left(\underline{\gamma}_2\right)$$
$$= (-1)^{\mu(x_0)+1+\maltese_1^{i-1}+\sigma(x_0,x_1,\ldots,x_d)}\epsilon_\partial\left(\underline{\gamma}_1, \underline{\gamma}_2\right) \quad \forall\left(\underline{\gamma}_1, \underline{\gamma}_2\right) \in \mathcal{A}_e \,.$$

Inserting this into (A.31) yields

$$(-1)^{\maltese_1^{i-1}}a_{\mathbf{Y}}^{d-l}(x_0, x_1, \ldots, x_{i-1}, y, x_{i+l+1}, \ldots, x_d) \cdot a_{\mathbf{Y}}^{l+1}(y, x_i, \ldots, x_{i+l})$$
$$= (-1)^{\sigma(x_0,x_1,\ldots,x_d)+\mu(x_0)+1}\sum_{T\in\mathrm{BinTree}_d}\sum_e\sum_{\left(\underline{\gamma}_1,\underline{\gamma}_2\right)}\epsilon_\partial\left(\underline{\gamma}_1, \underline{\gamma}_2\right) \cdot$$

where the sums are given as above. By taking the sum of the last equation over all i and l, i.e. over all types of internal edges, we derive the following result:

$$\sum_{i=1}^{d-1}\sum_{l=1}^{d-1-i}\sum_y(-1)^{\maltese_1^{i-1}}a_{\mathbf{Y}}^{d-l}(x_0, x_1, \ldots, x_{i-1}, y, x_{i+l+1}, \ldots, x_d) \cdot a_{\mathbf{Y}}^{l+1}(y, x_i, \ldots, x_{i+l})$$
$$= (-1)^{\sigma(x_0,x_1,\ldots,x_d)+\mu(x_0)+1}\sum_{T\in\mathrm{RTree}_d}\sum_{e\in E_{int}(T)}\sum_{\left(\underline{\gamma}_1,\underline{\gamma}_2\right)}\epsilon_\partial\left(\underline{\gamma}_1, \underline{\gamma}_2\right) \cdot \qquad (A.32)$$

We keep this equation in mind and continue by considering the other two types of boundary curves of $\mathcal{A}_{\mathbf{Y}}^d(x_0, x_1, \ldots, x_d)$, namely elements of spaces of type

$$\widehat{\mathcal{M}}(x_0, y_0) \times \mathcal{A}_{\mathbf{Y}}^d(y_0, x_1, \ldots, x_d, T) \quad \text{and}$$
$$\mathcal{A}_{\mathbf{Y}}^d(x_0, x_1, \ldots, x_{i-1}, y_i, x_{i+1}, \ldots, x_d, T) \times \widehat{\mathcal{M}}(y_i, x_i) \text{ for } i \in \{1, 2, \ldots, d\} \quad \text{and } T \in \mathrm{BinTree}_d \,.$$

We start with the former type. Let $y_0 \in \mathrm{Crit} f$ with $\mu(y_0) = \mu(x_0) - 1$. By definition, it holds that

$$a_{\mathbf{Y}}^1(x_0, y_0) \cdot a_{\mathbf{Y}}^d(y_0, x_1, \ldots, x_d)$$

$$= \sum_{T \in \mathrm{BinTree}_d} (-1)^{n+1+\sigma(y_0, x_1, \ldots, x_d)+r(T)} \#_{\mathrm{or}} \widehat{\mathcal{M}}(x_0, y_0) \cdot \#_{\mathrm{or}} \mathcal{A}_{\mathbf{Y}}^d(y_0, x_1, \ldots, x_d, T)$$

$$= \sum_{T \in \mathrm{BinTree}_d} \sum_{(\hat{\gamma}, \underline{\gamma}) \in \widehat{\mathcal{M}}(x_0, y_0) \times \mathcal{A}_{\mathbf{Y}}^d(y_0, x_1, \ldots, x_d, T)} (-1)^{\sigma(x_0, y_0)+\sigma(y_0, x_1 \ldots, x_d)+r(T)} \epsilon\left(\hat{\gamma}\right) \cdot \epsilon_T\left(\underline{\gamma}\right)$$

$$= \sum_{T \in \mathrm{BinTree}_d} \sum_{(\hat{\gamma}, \underline{\gamma}) \in \widehat{\mathcal{M}}(x_0, y_0) \times \mathcal{A}_{\mathbf{Y}}^d(y_0, x_1, \ldots, x_d, T)} (-1)^{\sigma(x_0, x_1 \ldots, x_d)+r(T)} \epsilon\left(\hat{\gamma}\right) \cdot \epsilon_T\left(\underline{\gamma}\right) ,$$

where we have used part 2 of Proposition A.31 in the last line. Part (a) of Theorem A.26 implies

$$\epsilon\left(\hat{\gamma}\right) \cdot \epsilon_T\left(\underline{\gamma}\right) = (-1)^{\mu(x_0)+1} \epsilon_{\partial, T}\left(\hat{\gamma}, \underline{\gamma}\right) = (-1)^{\mu(x_0)+1+r(T)} \epsilon_{\partial}\left(\hat{\gamma}, \underline{\gamma}\right)$$

for all $\left(\hat{\gamma}, \underline{\gamma}\right) \in \widehat{\mathcal{M}}(x_0, y_0) \times \mathcal{A}_{\mathbf{Y}}^d(y_0, x_1, \ldots, x_d, T)$ and $T \in \mathrm{BinTree}_d$. Inserting this into the above computation yields:

$$a_{\mathbf{Y}}^1(x_0, y_0) \cdot a_{\mathbf{Y}}^d(y_0, x_1, \ldots, x_d)$$
$$= (-1)^{\sigma(x_0, x_1 \ldots, x_d)+\mu(x_0)+1} \sum_{T \in \mathrm{BinTree}_d} \sum_{(\hat{\gamma}, \underline{\gamma}) \in \widehat{\mathcal{M}}(x_0, y_0) \times \mathcal{A}_{\mathbf{Y}}^d(y_0, x_1, \ldots, x_d, T)} \epsilon_{\partial}\left(\hat{\gamma}, \underline{\gamma}\right) . \quad \text{(A.33)}$$

Let $i \in \{1, 2, \ldots, d\}$ and $y_i \in \mathrm{Crit} f$ with $\mu(y_i) = \mu(x_i) + 1$. Then

$$a_{\mathbf{Y}}^d(x_0, x_1, \ldots, x_{i-1}, y_i, x_{i+1}, \ldots, x_d) \cdot a_{\mathbf{Y}}^1(y_i, x_i)$$
$$= \sum_{T \in \mathrm{BinTree}_d} \sum_{(\underline{\gamma}, \hat{\gamma}) \in \mathcal{A}_{\mathbf{Y}}^d(x_0, x_1, \ldots, x_{i-1}, y_i, x_{i+1}, \ldots, x_d, T) \times \widehat{\mathcal{M}}(y_i, x_i)} (-1)^{\sigma(x_0, x_1, \ldots, x_{i-1}, y_i, x_{i+1}, \ldots, x_d)+r(T)+\sigma(y_i, x_i)}$$

$$\epsilon\left(\underline{\gamma}\right) \cdot \epsilon\left(\hat{\gamma}\right) .$$

We derive from part (b) of Theorem A.26 that

$$\epsilon_T\left(\underline{\gamma}\right) \cdot \epsilon\left(\hat{\gamma}\right) = (-1)^{\mu(x_0)+1+(d-i)(n+1)+\maltese_1^{i-1}} \epsilon_{\partial, T}\left(\underline{\gamma}, \hat{\gamma}\right)$$
$$= (-1)^{\mu(x_0)+1+(d-i)(n+1)+\maltese_1^{i-1}+r(T)} \epsilon_{\partial}\left(\underline{\gamma}, \hat{\gamma}\right)$$

for all $\left(\underline{\gamma}, \hat{\gamma}\right) \in \mathcal{A}_{\mathbf{Y}}^d(x_0, x_1, \ldots, x_{i-1}, y_i, x_{i+1}, \ldots, x_d, T) \times \widehat{\mathcal{M}}(y_i, x_i)$ and $T \in \mathrm{BinTree}_d$. Inserting this into the above computation yields

$$a_{\mathbf{Y}}^d(x_0, x_1, \ldots, x_{i-1}, y_i, x_{i+1}, \ldots, x_d) \cdot a_{\mathbf{Y}}^1(y_i, x_i)$$
$$= (-1)^{\mu(x_0)+1+\sigma(x_0,x_1,\ldots,x_{i-1},y_i,x_{i+1},\ldots,x_d)+\sigma(y_i,x_i)+(d-i)(n+1)+\maltese_1^{i-1}}$$
$$\sum_{T \in \mathrm{BinTree}_d} \sum_{(\gamma,\hat{\gamma}) \in \mathcal{A}_{\mathbf{Y}}^d(x_0,x_1,\ldots,x_{i-1},y_i,x_{i+1},\ldots,x_d,T) \times \widehat{\mathcal{M}}(y_i,x_i)} \epsilon_\partial\left(\gamma, \hat{\gamma}\right).$$

By part 3 of Proposition A.31, this is equivalent to:

$$(-1)^{\maltese_1^{i-1}} a_{\mathbf{Y}}^d(x_0, x_1, \ldots, x_{i-1}, y_i, x_{i+1}, \ldots, x_d) \cdot a_{\mathbf{Y}}^1(y_i, x_i) \qquad (A.34)$$
$$= (-1)^{\sigma(x_0,x_1,\ldots,x_d)+\mu(x_0)+1} \sum_{T \in \mathrm{BinTree}_d} \sum_{(\gamma,\hat{\gamma}) \in \mathcal{A}_{\mathbf{Y}}^d(x_0,x_1,\ldots,x_{i-1},y_i,x_{i+1},\ldots,x_d,T) \times \widehat{\mathcal{M}}(y_i,x_i)} \epsilon_\partial\left(\gamma, \hat{\gamma}\right).$$

Finally, we combine Eqs. (A.32)–(A.34) to prove the statement:

$$\sum_{i=1}^{d}\sum_{l=0}^{d-i}\sum_{y}(-1)^{\maltese_1^{i-1}} a_{\mathbf{Y}}^{d-l}(x_0,x_1,\ldots,x_{i-1},y,x_{i+l+1},\ldots,x_d) \cdot a_{\mathbf{Y}}^{l+1}(y,x_i,\ldots,x_{i+l})$$
$$= \sum_{\substack{y_0 \in \mathrm{Crit}f \\ \mu(y_0)=\mu(x_0)-1}} a_{\mathbf{Y}}^1(x_0,y_0) \cdot a_{\mathbf{Y}}^d(y_0,x_1,\ldots,x_d)$$
$$+ \sum_{i=1}^{d-1}\sum_{l=1}^{d-1-i}\sum_{y}(-1)^{\maltese_1^{i-1}} a_{\mathbf{Y}}^{d-l}(x_0,x_1,\ldots,x_{i-1},y,x_{i+l+1},\ldots,x_d) \cdot a_{\mathbf{Y}}^{l+1}(y,x_i,\ldots,x_{i+l})$$
$$+ \sum_{i=1}^{d}\sum_{\substack{y_i \in \mathrm{Crit}f \\ \mu(y_i)=\mu(x_i)+1}}(-1)^{\maltese_1^{i-1}} a_{\mathbf{Y}}^d(x_0,x_1,\ldots,x_{i-1},y_i,x_{i+1},\ldots,x_d) \cdot a_{\mathbf{Y}}^1(y_i,x_i)$$
$$= (-1)^{\sigma(x_0,x_1\ldots,x_d)+\mu(x_0)+1} \sum_{\mu(y_0)=\mu(x_0)-1}\sum_{T \in \mathrm{BinTree}_d}\sum_{(\hat{\gamma},\gamma) \in \widehat{\mathcal{M}}(x_0,y_0) \times \mathcal{A}_{\mathbf{Y}}^d(y_0,x_1,\ldots,x_d,T)} \epsilon_\partial\left(\hat{\gamma}, \gamma\right)$$
$$+ (-1)^{\sigma(x_0,x_1,\ldots,x_d)+\mu(x_0)+1} \sum_{T \in \mathrm{BinTree}_d}\sum_{e \in E_{int}(T)}\sum_{(\gamma_1,\gamma_2)} \epsilon_\partial\left(\gamma_1, \gamma_2\right)$$
$$+ (-1)^{\sigma(x_0,\ldots,x_d)+\mu(x_0)+1} \sum_{i=1}^{d}\sum_{\mu(y_i)=\mu(x_i)+1}\sum_{T \in \mathrm{BinTree}_d}\sum_{(\gamma,\hat{\gamma}) \in \mathcal{A}_{\mathbf{Y}}^d(x_0,x_1,\ldots,y_i,\ldots,x_d,T) \times \widehat{\mathcal{M}}(y_i,x_i)} \epsilon_\partial\left(\gamma, \hat{\gamma}\right)$$
$$= (-1)^{\sigma(x_0,x_1\ldots,x_d)+\mu(x_0)+1} \sum_{(\gamma_1,\gamma_2) \in \partial\overline{\mathcal{A}}_{\mathbf{Y}}^d(x_0,x_1,\ldots,x_d)} \epsilon_\partial\left(\gamma_1, \gamma_2\right) = 0,$$

where we used (A.14) and the boundary description from p.99 for the last equality. $\qquad\square$

References

[AB95]   Austin, D.M., Braam, P.J.: Morse-Bott theory and equivariant cohomology. The Floer Memorial Volume. Progress in Mathematics, vol. 133, pp. 123–183. Birkhäuser, Basel (1995)

[Abo09]   Abouzaid, M.: Morse homology, tropical geometry, and homological mirror symmetry for toric varieties. Selecta Math. (N.S.) 15(2), 189–270 (2009)

[Abo10]   Abouzaid, M.: A geometric criterion for generating the Fukaya category. Publ. Math. Inst. Hautes Études Sci. 112, 191–240 (2010)

[Abo11]   Abouzaid, M.: A topological model for the Fukaya categories of plumbings. J. Differ. Geom. 87(1), 1–80 (2011)

[Abo12]   Abouzaid, M.: Framed bordism and Lagrangian embeddings of exotic spheres. Ann. Math. (2) 175(1), 71–185 (2012)

[AD14]   Audin, M., Damian, M.: Morse Theory and Floer Homology. Universitext. Springer, London; EDP Sciences, Les Ulis (2014); Translated from the 2010 French original by Reinie Erné

[AK16]   Abouzaid, M., Kragh, T.: Simple homotopy equivalence of nearby Lagrangians (2016). arXiv:1603.05431

[Aka07]   Akaho, M.: Morse homology and manifolds with boundary. Commun. Contemp. Math. 9(3), 301–334 (2007)

[AM06]   Abbondandolo, A., Majer, P.: Lectures on the Morse complex for infinite-dimensional manifolds. Morse-Theoretic Methods in Nonlinear Analysis and in Symplectic Topology. Nato Science Series II: - Mathematics, Physics and Chemistry, vol. 217, pp. 1–74. Springer, Dordrecht (2006)

[AR67]   Abraham, R., Robbin, J.: Transversal Mappings and Flows. W. A. Benjamin Inc, New York-Amsterdam (1967)

[AS10]   Abbondandolo, A., Schwarz, M.: Floer homology of cotangent bundles and the loop product. Geom. Topol. 14(3), 1569–1722 (2010)

[Aur14]   Auroux, D.: A beginner's introduction to Fukaya categories. Contact and Symplectic Topology. Bolyai Society Mathematical Studies, vol. 26, pp. 85–136. János Bolyai Mathematical Society, Budapest (2014)

[BH04]   Banyaga, A., Hurtubise, D.: Lectures on Morse Homology. Kluwer Texts in the Mathematical Sciences, vol. 29. Kluwer Academic Publishers Group, Dordrecht (2004)

[CJ02]   Cohen, R.L., Jones, J.D.S.: A homotopy theoretic realization of string topology. Math. Ann. 324(4), 773–798 (2002)

© Springer International Publishing AG, part of Springer Nature 2018
S. Mescher, *Perturbed Gradient Flow Trees and A∞-algebra Structures in Morse Cohomology*, Atlantis Studies in Dynamical Systems 6,
https://doi.org/10.1007/978-3-319-76584-6

[CN12]   Cohen, R.L., Norbury, P.: Morse field theory. Asian J. Math. **16**(4), 661–712 (2012)

[CO15]   Cieliebak, K., Oancea, A.: Symplectic homology and the Eilenberg-Steenrod axioms (2015). arXiv:1511.00485

[CS99]   Chas, M., Sullivan, D.: String topology (1999). arXiv:math/9911159

[CS09]   Cohen, R.L., Schwarz, M.: A Morse theoretic description of string topology. New Perspectives and Challenges in Symplectic Field Theory. CRM Proceedings and Lecture Notes, vol. 49, pp. 147–172. American Mathematical Society, Providence (2009)

[CV06]   Cohen, R.L., Voronov, A.A.: Notes on string topology. String Topology and Cyclic Homology. Advanced Courses in Mathematics. CRM Barcelona, pp. 1–95. Birkhäuser, Basel (2006)

[FH93]   Floer, A., Hofer, H.: Coherent orientations for periodic orbit problems in symplectic geometry. Math. Z. **212**(1), 13–38 (1993)

[FHS95]   Floer, A., Hofer, H., Salamon, D.: Transversality in elliptic Morse theory for the symplectic action. Duke Math. J. **80**(1), 251–292 (1995)

[Flo88a]   Floer, A.: Morse theory for Lagrangian intersections. J. Differ. Geom. **28**(3), 513–547 (1988)

[Flo88b]   Floer, A.: A relative Morse index for the symplectic action. Comm. Pure Appl. Math. **41**(4), 393–407 (1988)

[Flo88c]   Floer, A.: The unregularized gradient flow of the symplectic action. Comm. Pure Appl. Math. **41**(6), 775–813 (1988)

[Flo89a]   Floer, A.: Symplectic fixed points and holomorphic spheres. Comm. Math. Phys. **120**(4), 575–611 (1989)

[Flo89b]   Floer, A.: Witten's complex and infinite-dimensional Morse theory. J. Differ. Geom. **30**(1), 207–221 (1989)

[FO97]   Fukaya, K., Oh, Y.-G.: Zero-loop open strings in the cotangent bundle and Morse homotopy. Asian J. Math. **1**(1), 96–180 (1997)

[FO99]   Fukaya, K., Ono, K.: Arnold conjecture and Gromov-Witten invariant. Topology **38**(5), 933–1048 (1999)

[FOOO09]   Fukaya, K., Oh, Y.-G., Ohta, H., Ono, K.: Lagrangian Intersection Floer Theory: Anomaly and Obstruction. Parts I and II. AMS/IP Studies in Advanced Mathematics, vol. 46. American Mathematical Society, Providence; International Press, Somerville (2009)

[Fuk93]   Fukaya, K.: Morse homotopy, A^{∞}-category, and Floer homologies. In: Proceedings of GARC Workshop on Geometry and Topology '93 (Seoul, 1993) (Seoul). Lecture Notes Series, vol. 18, pp. 1–102. Seoul National University (1993)

[Fuk97]   Fukaya, K.: Morse homotopy and its quantization. Geometric Topology (Athens, GA, 1993). AMS/IP Studies in Advanced Mathematics, vol. 2, pp. 409–440. American Mathematical Society, Providence (1997)

[GJ90]   Getzler, E., Jones, J.D.S.: A_{∞}-algebras and the cyclic bar complex. Illinois J. Math. **34**(2), 256–283 (1990)

[GP74]   Guillemin, V., Pollack, A.: Differential Topology. Prentice-Hall Inc., Englewood Cliffs (1974)

[Hir76]   Hirsch, M.W.: Differential Topology. Graduate Texts in Mathematics, vol. 33. Springer, New York (1976)

[Hut02]   Hutchings, M.: Lecture notes on Morse homology (with an eye towards Floer theory and pseudoholomorphic curves), UC Berkeley (2002). http://math.berkeley.edu/~hutching/teach/276-2010/mfp.ps

[HWZ07]   Hofer, H., Wysocki, K., Zehnder, E.: A general Fredholm theory. I. A splicing-based differential geometry. J. Eur. Math. Soc. **9**(4), 841–876 (2007)

[Irw80]   Irwin, M.C.: Smooth Dynamical Systems. Pure and Applied Mathematics, vol. 94. Academic Press Inc. [Harcourt Brace Jovanovich Publishers], New York (1980)

[Jon87]   Jones, J.D.S.: Cyclic homology and equivariant homology. Invent. Math. **87**(2), 403–423 (1987)

[Jos08] Jost, J.: Riemannian Geometry and Geometric Analysis. Universitext, 5th edn. Springer, Berlin (2008)

[KM07] Kronheimer, P., Mrowka, T.: Monopoles and Three-Manifolds. New Mathematical Monographs, vol. 10. Cambridge University Press, Cambridge (2007)

[Kon95] Kontsevich, M.: Homological algebra of mirror symmetry. In: Proceedings of the International Congress of Mathematicians, vol. 1, 2, pp. 120–139. Birkhäuser, Basel (1995). (Zürich, 1994)

[KS10] Kreck, M., Singhof, W.: Homology and cohomology theories on manifolds. Münster J. Math. 3, 1–9 (2010)

[Lau11] Laudenbach, F.: A Morse complex on manifolds with boundary. Geom. Dedicata 153, 47–57 (2011)

[Lee03] Lee, J.M.: Introduction to Smooth Manifolds. Graduate Texts in Mathematics, vol. 218. Springer, New York (2003)

[Lod98] Loday, J.-L.: Cyclic Homology. Grundlehren der Mathematischen Wissenschaften, vol. 301, 2nd edn. Springer, Berlin (1998)

[LT98] Liu, G., Tian, G.: Floer homology and Arnold conjecture. J. Differ. Geom. 49(1), 1–74 (1998)

[Mar92] Markl, M.: A cohomology theory for $A(m)$-algebras and applications. J. Pure Appl. Algebr. 83(2), 141–175 (1992)

[Mat02] Matsumoto, Y.: An Introduction to Morse Theory. Translations of Mathematical Monographs, vol. 208. American Mathematical Society, Providence (2002)

[Mes16a] Mescher, S.: Hochschild homology of Morse cochain complexes and free loop spaces, Ph.D. thesis, Universität Leipzig (2016)

[Mes16b] Mescher, S.: A primer on A_∞-algebras and their Hochschild homology (2016). arXiv:1601.03963

[Mil63] Milnor, J.W.: Morse Theory. Based on lecture notes by M. Spivak and R. Wells. Annals of Mathematics Studies, No. 51. Princeton University Press, Princeton (1963)

[Nic11] Nicolaescu, L.: An Invitation to Morse Theory. Universitext, 2nd edn. Springer, New York (2011)

[Sch93] Schwarz, M.: Morse Homology. Progress in Mathematics, vol. 111. Birkhäuser, Basel (1993)

[Sch99] Schwarz, M.: Equivalences for Morse homology. Geometry and Topology in Dynamics. Contemporary Mathematics, vol. 246, pp. 197–216. American Mathematical Society, Providence (1999)

[Sch05] Schätz, F.: The Morse-Smale complex, Diplomarbeit (Master's Thesis), University of Vienna (2005)

[Sch12] Schneider, M.: The Leray-Serre spectral sequence in Morse homology on Hilbert manifolds and in Floer homology on cotangent bundles, Ph.D. thesis, Universität Leipzig (2012)

[Sco05] Scorpan, A.: The Wild World of 4-Manifolds. American Mathematical Society, Providence (2005)

[Sei08] Seidel, P.: Fukaya Categories and Picard-Lefschetz Theory. European Mathematical Society (EMS), Zürich, Zurich Lectures in Advanced Mathematics (2008)

[Sma61] Smale, S.: On gradient dynamical systems. Ann. of Math. 2(74), 199–206 (1961)

[Sta63a] Stasheff, J.D.: Homotopy associativity of H-spaces. I. Trans. Amer. Math. Soc. 108, 275–292 (1963)

[Sta63b] Stasheff, J.D.: Homotopy associativity of H-spaces. II. Trans. Amer. Math. Soc. 108, 293–312 (1963)

[Tra08a] Tradler, T.: The Batalin-Vilkovisky algebra on Hochschild cohomology induced by infinity inner products. Ann. Inst. Fourier (Grenoble) 58(7), 2351–2379 (2008)

[Tra08b] Tradler, T.: Infinity-inner-products on A-infinity-algebras. J. Homotopy Relat. Struct. 3(1), 245–271 (2008)

[Voi14] Voigt, R.: Transport functions and Morse K-theory, Ph.D. thesis, Universität Leipzig (2014)

[Web06]   Weber, J.: The Morse-Witten complex via dynamical systems. Expo. Math. **24**(2), 127–159 (2006)
[Weh12]   Wehrheim, K.: Smooth structures on Morse trajectory spaces, featuring finite ends and associative gluing. In: Proceedings of the Freedman Fest. Geometry and Topology Monographs, vol. 18, pp. 369–450. Geom. Topol. Publ., Coventry (2012)
[Wei94]   Weibel, C.A.: An Introduction to Homological Algebra. Cambridge Studies in Advanced Mathematics, vol. 38. Cambridge University Press, Cambridge (1994)

Index

Printed in the United States
By Bookmasters